Leadership Lessons from the Military

Leadership Lessons from the Military

Dheeraj Sharma

www.sagepublications.com
Los Angeles • London • New Delhi • Singapore • Washington DC

Copyright © Dheeraj Sharma, 2014

All rights reserved. No part of this book may be reproduced or utilized in any form or by any means, electronic or mechanical, including photocopying, recording or by any information storage or retrieval system, without permission in writing from the publisher.

First published in 2014 by

SAGE Response
B1/I-1 Mohan Cooperative Industrial Area
Mathura Road, New Delhi 110 044, India

SAGE Publications Inc
2455 Teller Road
Thousand Oaks, California 91320, USA

SAGE Publications Ltd
1 Oliver's Yard, 55 City Road
London EC1Y 1SP, United Kingdom

SAGE Publications Asia-Pacific Pte Ltd
3 Church Street
#10-04 Samsung Hub
Singapore 049483

Published by Vivek Mehra for SAGE Publications India Pvt Ltd, typeset at 11/13pt Bembo by Diligent Typesetter, Delhi and printed at Sai Print-o-Pack, New Delhi.

Library of Congress Cataloging-in-Publication Data

Sharma, Dheeraj.
  Leadership lessons from the military / Dheeraj Sharma.
  pages cm
  Includes bibliographical references.
  1. Leadership. 2. Command of troops. 3. Military art and science. I. Title.
  HD57.7.S4757   658.4'092—dc23   2014   2014019764

ISBN: 978-81-321-1848-0 (PB)

The SAGE Team: Sachin Sharma, Neha Sharma, Rajib Chatterjee, and Rajinder Kaur

*For my parents Pran Nath and Sangeeta Sharma,
my wife Shveta Sharma, and my children Lavya and Girik*

Thank you for choosing a SAGE product! If you have any comment, observation or feedback, I would like to personally hear from you. Please write to me at contactceo@sagepub.in

—Vivek Mehra, Managing Director and CEO,
SAGE Publications India Pvt Ltd, New Delhi

Bulk Sales

SAGE India offers special discounts for purchase of books in bulk. We also make available special imprints and excerpts from our books on demand.

For orders and enquiries, write to us at

Marketing Department
SAGE Publications India Pvt Ltd
B1/I-1, Mohan Cooperative Industrial Area
Mathura Road, Post Bag 7
New Delhi 110044, India
E-mail us at marketing@sagepub.in

Get to know more about SAGE, be invited to SAGE events, get on our mailing list. Write today to marketing@sagepub.in

This book is also available as an e-book.

CONTENTS

List of Figures   ix
Foreword by Lt General (Retd.) Sanjeev Anand   xi
Acknowledgments   xiii

1. Introduction   1
2. Lessons of Leadership   20
3. Building Teams: Creating a Person–Organization Fit   47
4. Workforce Motivation   70
5. Organizational Climate and Culture: Creating a Climate of Trust   93
6. Developing SOPs: Strategies and Tactics   114
7. Work–Life Balance   134
8. Organizational Pride and Unity   161

About the Author   184

LIST OF FIGURES

3.1. Stages of Team Formation   59

4.1. Antecedents of Pre-Training Motivation   74

5.1. Social-Learning Constructs   108

6.1. Similarities and Differences between the Military and
     Corporate Firms   115

7.1. Factors Influenced by Family-Friendly Environment   139

8.1. The Ashridge's Model Mission for Organizations   166

FOREWORD

We are kept from our goal, not by obstacles, but by a clear path to a lesser goal.

—The Bhagavad Gita

It is my pleasure and honor to provide a foreword to this outstanding book, written by Professor Dheeraj Sharma of Indian Institute of Management, Ahmedabad. Notably, the uniqueness of this book is attributed to the creation of an amalgam of military practices and management techniques. The book deals with the development of corporate capabilities based on military practices. It offers a unique opportunity to touch a wider audience and provide fundamental information of military practices, which may be relevant to the corporate world. This book is written by the author who has been deeply involved in the military processes for the past two years, and who strives to find innovative solutions to some of the complex management problems in this field, thus, resulting in the utmost level of authenticity.

Military approaches and processes are complex issues, and, sometimes, less understood. Military thrives on great leadership. However, the biggest lesson that is learnt from the military is to learn to follow. The military provides continuous education and training on leadership and followership. Through various training programs, regimental traditions, and mentorship, the military stresses a wide range of styles, methods, and techniques to tackle organizational challenges. Most of the practices in the military are geared toward influencing an individual to engage in an appropriate behavior, and creating an ecosystem that fosters goal attainment.

Over the years, there might have been two major, but contrary, views that have explained performance in the corporate world. First, corporate policies may be more prominent than the values of the employees and the management. Second, according to the contrary view, the character and values of the employees and the management form the basis of better firm performance. The book reconciles the two contrary views by suggesting that good corporate performance is derived from military practices and approaches, which instill ethical and innovative behavior on one hand, and create standard operating procedures and drills on the other hand.

The book provides a comprehensive overview of the approaches and processes of the military in managing organizational challenges. In particular, the book offers an insight into the military approach to leadership, workforce motivation, building and managing teams, creating an ethical environment that foster ethical conduct, creating standard operating procedures, creating work–life balance, and developing and nurturing a vibrant organizational culture. Overall, the book helps in understanding that organizational outcomes are not necessarily due to the firm, but they can be attributed to individual leadership as well.

This book provides sufficient evidence that military practices and approaches are transferable to corporate set-ups. Consequently, I have no doubt that this book will foster discussion not only in the corporate world, but also in the military circuit. I am sanguine about the fact that it will contribute to a better and more coherent development of general management capabilities. Moreover, I am of the view that it will give new impetus to closer integration of general management principles with military principles of leadership.

Lt General (Retd.) Sanjeev Anand
PVSM, AVSM, VSM
Former Adjutant General,
Indian Army

ACKNOWLEDGMENTS

I thought of writing this book after I became closely associated with Army Training Command through Army Management Studies Board (AMSB). AMSB is platform for exchange between top brass of Indian Army and management researchers on innovative ideas and issues that deal with organisational effectiveness and national security. I was tasked by AMSB with a project that required me to review and recommend officer selection process for Indian Army. During the course of my project, I interacted with over a hundred army officers and several hundred cadets. During my interactions, I learned that corporate world could benefit considerably from learning some of the practices, procedures, and processes of Indian army.

This book was written for corporate manager and army officers who are hoping to pursue a corporate career after serving in the army. This book was developed based on my research and understanding of practices of armed forces.

Consequently, I offers my deepest appreciation to all those who have contributed to this book. What was accomplished was the result of our joint efforts. However, special appreciation is extended to the officers of Indian army, who were instrumental in coordinating and managing my interactions.

The management and coordination of my interactions requires dedication and enormous effort of officers of Indian Military Academy. Indian Military Academy played pivotal role in encouraging participation of various formations and bodies for me to complete my research project conscientiously.

I want thank my team at IIM-Ahmedabad—Divya, Kirti, Shivraj, Sonu Kumar, Tanvi, my IIT interns and many others for their dedicated work. I want to extend my gratitude to Lt General Sanjeev Madhok (Army Commander-ARTRAC), Lt General Rajeev Anand (Adjutant General), and Lt General Balbir Pama (DG-Recruiting) for their wonderful insight. Also, I want to extend special thanks to Lt General Manvendra Singh, Commandant-Indian Military Academy for his remarkable support in managing various administrative tasks related to my project and providing special insights for writing this book.

Additionally, I want to extend my gratitude to my family for their time and patience when I was authoring this book and I really appreciate it. Finally, through this book I honor, salute and celebrate our troops and their families, who have been instrumental in creating the ethos, culture, and value system of Indian army. Jai Hind!

1
INTRODUCTION

There is so much that the corporate sector can learn from the military; right from laying the base of the organization to fighting competition with rivals, and the trajectory to being on the top. The genesis of the military goes back to eons. Military has not been studied as a part of management studies in general. However, the fact is that management, as a subject, or even as a concept, is based on the military. Militaries are undisputedly the best managed organizations in the world. Although it has a unique model, and it follows unique operations and processes, there are lessons that the corporate organizations can derive from it. If one can see beyond the abstruse, one will be able to see the similarities between the corporate and the military. There are various strategies and lessons that the corporate organizations can learn and adapt from the military.

In the seventeenth century, the concept of military science as a separate branch of knowledge did not exist. The officers' corps in that century were considered as aristocrats, rather than effective performers of military functions. Their appointment then was based on their wealth, birth, personal and political influence, and not on the basis of competence. Military command was considered to be an art form, such as music or drawing, which could only be acquired through inheritance. Military aptitude was considered a purely subjective concept, according to which core military competence laid within the individuals could be transmitted to or learned from the environment. There was no existence of a professional body of knowledge, except for a few technical schools to impart military knowledge. There was even an absence of any system for applying that knowledge in practice. The whole concept of the military was based on the notion that certain few men are born to command and others are born to obey, and that the man born with superior native abilities could be a successful commander. The fact that there was a possibility of developing military officers or generals through social institutions was beyond the imagination for these thinkers.

In the eighteenth century, new developments in the concept of military were seen. This was a new beginning, though faint, to the modern outlook on military. There was an emergence of few writers who contributed to the military concept through their noteworthy work on specialized topics, and even anticipated the developments that would take place in the nineteenth century. But their views were based on the past military practices of the Greeks and the Romans. Moreover, their views were devoid of the reality aspect and, thus, had several limitations.

Professionalization of the military was said to be have concentrated in two periods of the nineteenth century. During Napoleonic and post-Napoleonic war, most nations developed institutions for military education and liberalized the entry restrictions for the officers' corps. The third quarter of the century saw more advanced changes in the form of overhauling of promotion, managing the general staff, and the establishment of more advanced military educational institutes. The military started being treated as a profession during this century, and the nations had started furnishing resources to support the military as a permanent profession.[1] Since then, the military has not looked back and has continued to see many changes in both technology and in infrastructure. Although, the military is treated as a separate profession, it is different from other professions in several

[1] Huntington, S. P. (1985). *The soldier and the state: The theory and politics of civil military relation*. Cambridge, Massachusetts, United States of America: Belknap Press of Harvard University Press.

aspects, such as the attitude of its employees, which is self-sacrificing, and the attitude of "do or die." Further, the physical courage they have is not possessed by any other employees from any profession. Military organization is one single company comprising of army, air force, and navy. So, once an individual is committed to one of its wings, he/she cannot move out from that wing to some other wing, unlike in other professions, where an individual can move from one company to another.[2]

ROLE OF THE MILITARY

The military has been playing an important role in maintaining peace and harmony across the world. It has helped in developing the economies of countries and the world. The major issues that the military personnel face are the nature of tenure, length of service, and the extent of flexibility it offers to cope up with the changing needs of leadership over a period of time.[3]

The military can be defined as

> a conglomerate of people, who come together from different culture and environment with the purpose of performing military tasks. These units work, live, eat, and sleep together to the extent that they become a one big "family." This family is so close to each other that they know each other's strengths and weaknesses, likes and dislikes, their beliefs and ideas.[4]

The military delivers its services through the people and for the people. The military processes involve the optimum utilization of limited resources available at its disposal, including soft and hard services, using intellectual capital. It relies on the operationalized

---

[2] Horvath, J. A., Forsythe, B. B., & Bullis, R.C. (1999). *Tacit knowledge in professional practice: Researcher and practitioner perspectives*. Mahwah, NJ: Lawrence Erlbaum Associates.

[3] Singh, J. (June 7, 2004). Leadership for the 21st century. *The Indian Express*. Retrieved from http://www.indianexpress.com/oldStory/48442/ (accessed on July 31, 2014).

[4] Field manual (FM-22-102). Soldier team development. Characteristics of a combat ready team.

and formalized processes, the delivery of which depends upon the effective chain of communication.[5]

LEADERSHIP: IN THE EARLIER TIMES

Leadership has had its roots since the inception of civilization. Throughout the ages, there have been people whose contribution to history is unforgettable in terms of their tremendous influence on people's thought processes and actions. These people were leaders but given different names in ancient times, such as *head of state*, *military commander*, *king*, *chief*, etc.

It is said that an organization without a leader would lead to a drop in its smooth and easy functionality, and finally leads to stagnation. A leader is not only the guardian of its members, but also a companion to them. He/she is the person whom people turn to when they face a problem. The leader plans the strategies, motivates the subordinates to work for the target in hand, and is responsible for its perfect execution. He/she places the company and his/her people's welfare above his/her own interests. He/she has an acute vision and a knack for problem solving. Napoleon once said, "An army of rabbits led by a lion is much better than an army of lions led by a rabbit."[6]

Over a period of time, leadership has evolved from an authoritative style to a style that provides a more comfortable environment, and then, finally, to a style which empowers and encourages the employees, and provides assistance in their personal and organizational growth and development. Although there are many definitions of leadership, the term "leadership" can be explained as an effort and power to induce compliance.[7]

---

[5] Hinks, J., Alexander, M., & Dunlop, G. (2007). Translating military experiences of managing innovation and innovativeness into FM. *Journal of Facilities Management*, 5(4), 226–242.

[6] Muniapan, B., & Dass, M. (2009). An Indian leadership perspective from literature works of Poet Kannadasan. *International Journal of Indian Culture and Business Management*, 2(3), 326–340.

[7] Stone, A. G., & Patterson, K. (2005). *History of leadership focus*. School of Leadership Studies, Regent University. Retrieved from http://www.regent.edu/acad/global/publications/sl_proceedings/2005/stone_history.pdf (accessed on July 31, 2014).

EVOLUTION OF LEADERSHIP FROM TRAITS TO SITUATIONAL FACTORS

The introductory phase in leadership was marked by laying focus on the *personality traits* of the leader. It proposed that leadership can be developed by absorbing the qualities of an effective leader.[8] However, this theory had some limitations. The believers of this theory could not identify the common personality traits of all the leaders. Further, the fact that individuals differ in their ability to learn these traits was not considered. The whole concept, thus, failed, as it was a mere theoretical framework and had no practical implications. This gave rise to the next phase in the leadership history.

Believers of the new phase based their leadership theories on military leaders and powerful kings, who had the ability to lead thousands of soldiers in a war. They viewed leadership as the *ability to persuade* others to a specific course-of-action. However, this theory, though applicable today, is not found to be effective.

The next phase was an extension of the trait theory that gave prominence to *human behavior* as the basis for leadership. The theorists supporting this view concluded that leadership is not concerned with influencing people to do the required task. Rather, it is concerned with shaping the behavior of individuals through stimulation in the form of modification in the personal behavior of the leader.

The next stage in leadership gave importance to *situational factors* as the means to define leadership. The believers in this phase believed that leadership is based on the task, status of the leader, and the nature of the external situations.

EVOLUTION OF LEADERSHIP: FROM CONTINGENCY TO TRANSFORMATIONAL

At this stage, many researchers found that all the four factors, that is, personality traits, persuasive power, behavioral patterns, and situational factors threw light on the requirements of being a

---

[8] Anonymous. (August 15, 2010).

good leader. However, it did not provide a clear idea about leadership, which led to the evolution of the concept of transactional leadership.

The supporters of transactional leadership believed that in this form of leadership, the leader is in continuous interaction with his followers, explaining them what has to be done and how to do it. They were also informed about the benefits that accrue on completion of the tasks and the penalty that would be levied for the non-completion of the tasks. However, this type of leadership seems to have lost its relevance in the recent times because it is believed that using coercive and legitimate powers to get things done is not an effective tool to motivate the employees and ensure their commitment.[9]

The limitation of this theory led to the next stage, where the macro views of leadership were viwed as leadership beyond individuals and groups. This was an era where the cultural aspect came into consideration. Leadership was seen to exist in the cultural context, rather than in any particular individual. The cultural processes were responsible for producing leaders in the future. Only when the culture had to be changed, a formal leadership outside the cultural framework was required.

Then came the present era, where transformational leadership is considered to be an effective style to get things done through others. According to this view, transformational leadership occurs when the leader does not lead the followers through the power of authority, but motivates them to perform tasks for the accomplishment of the group goals. The leader is responsible to define the mission, get it accepted from his/her followers, generate interest in the mission among the followers, and motivate them to overlook their personal goals over accomplishing the group goals.[10]

Leadership evolution has passed through various stages over the years—from personality traits to situational and from transactional to transformational. In spite of the differences in the opinion of the

---

[9] Bass, B. M. (1990). From transactional to transformational leadership: Learning to share the vision. *Organizational Dynamics, 18*(3), 19–31.
[10] Ibid.

researchers in all the stages, the traces of these views are seen to be present in the current scenario of leadership.

LEADERSHIP VERSUS MANAGEMENT

Many academicians and researchers have been debating on the relationship between leadership and management. Two alternative concepts have emerged with respect to their relationship. The two concepts are either different or they overlap.

Leadership and Management: Two Different Concepts

The first view believes that leadership and management are two mutually exclusive concepts. The believers of this concept often speak of terms such as "leaders" and "managers," rather than "leadership" and "management." The supporters of this belief differentiated between leaders and managers in terms of their motivation, personal history, way of thinking, and action. They view managers as problem solvers and leaders as visionaries. Further, managers have a narrow perspective and work toward developing skills required to perform the routine tasks, while leaders have a broad perspective and work toward forecasting and implementing the future organizational needs, and to implement changes when required.

Leadership and Management: Overlapping Concepts

The believers of this concept feel that the two concepts overlap each other to fulfill the expectations of the organizational roles. The supporters believe that one of the functions of the manager is to be a leader. It is believed that an individual needs to be both a leader and a manager to perform the functions of a supervisor. The researchers are of the belief that the leaders should manage and the managers should lead in order to meet their role expectations.

In the context of military, overlapping position of management and leadership is accepted. The term "leader" is used to refer to all the individuals on the post of supervisors. The term "leader" is used as a role model in the military in the same way as the term manager is used as a role model in the civil organizations. Thus, in the military, the term "leadership" represents a set of processes that involves exercising influence on others to accomplish the organizational goals.[11]

MILITARY LEADERSHIP

The military is the finest and most well-organized institution of the world. All the army men have their duties and tasks clearly mentioned. Since mistakes are not allowed, the army is led and served by able leaders. Leadership is the only aspect that separates the military from the rest of the organizations. Military leadership is defined by The Army Regulation (AR 600-100) as "influencing people by providing purpose, direction, and motivation, while operating to accomplish the mission and improve the organization."[12]

The military bestows the attitude of foregoing the personal interest in individuals in favor of the interest of the organization and the colleagues. This requires a proper alignment of the individual needs with the organizational needs, so that the purpose of both the parties is met.

Military leadership is an agreement between the leader and the followers, where the leader is required to be highly knowledgeable, courageous, and spontaneous in decision making, and willing to take the responsibility for the work done, and the followers are able to develop a feeling of belonging for the leader through the act he has performed.[13]

---

[11] Hovrath, J. A., William, W. M., Forsythe, G. B., Sweeney, P. J., Sternberg, R. J., McNally, J. A., & Wattendorf, J. (1994). *Tacit knowledge in military leadership: A review of the literature* (Technical Report No. 1017). Alexandria, Virginia, United States of America: United States Army Research Institute for the Behavioral and Social Sciences.

[12] Chan, C. C. (2003). The strategy process: A military-business comparison. *Leadership & Organization Development Journal, 24*(5), 304–305.

[13] Grattan, R. F. (2002). *The strategy process: A military business comparison*. Chippenham, Wiltshire, GB: Antony Rowe Ltd.

MILITARY LEADERSHIP VERSUS CORPORATE LEADERSHIP

Leadership is both different and similar in some aspects in the military and the corporate. Military leadership is different from corporate leadership in the sense that leadership in the military is considered as a principle that is inculcated throughout the system. Whereas, in the *corporate context*, management is considered as a core function and leadership is treated as a part of it. Leadership in the corporate sector is exercised only when the situation demands.

Military leadership is similar to corporate leadership in various ways. When individuals stay inside the boundaries of the military, they are safe, secured, and protected. But when they leave the boundary, they leave the comfortable environment, and are surrounded by an environment characterized by uncertainty and danger. In such situations, a leader is required to render support, guidance, and direction to the followers to make them calm and relaxed in situations with high tension. Similarly, when the individuals in the corporate leave the boundary of their organization, they leave the certain and predictable environment of the organization, and enter an environment which serves as a testing zone for them to test their courage in leadership.[14]

An army leader has the ability and the qualities to lead many corporate organizations at the same time. Brought up in an environment where decisions have to be quick and decisions can be life threatening, it becomes much easier for them to be a CEO. An army officer starts learning and imbibing leadership traits from a very junior level, and these skills get honed up by the time he/she becomes a brigadier. There have been leaders, both in the history and in the recent past, who were able military leaders and had enough in them to lead an army of thousands of men. They were not only respected as leaders, but also as capable companions who led from the front. None of these leaders had the same style of leadership, which made them stand out, but they all had a few basic traits that separate a leader of

---

[14] Voyer, P. (2011). *Courage in leadership*. Retrieved from http://www.iveybusinessjournal.com/topics/leadership/courage-in-leadership-from-the-battlefield-to-the-boardroom (accessed on July 31, 2014).

the classes from the leader of the masses. Irrespective of the place or era that they belonged to, they had both the physical and the moral courage to handle any condition calmly, an intelligent mind to go about a problem, and a ruthless attitude if any condition needs to be fulfilled. There have been many examples of leaders who dedicated their lives to the purpose and seldom cared about their fate. These leaders, whether young or old, have a lot in their stories to learn from. The corporate world might not be as testing and enduring as the army; but the army personnel can be treated as benchmarks for the corporate leaders because a military leader does not have to be just good, he has to be perfect.

MILITARY ASPECTS ABSENT IN CORPORATE ORGANIZATIONS[15]

- **Developing Junior Leaders**
  To develop the judgmental skills of the junior leaders in the military, they are empowered to take initiatives and flexibility in making decisions through decentralized execution. This gives the leaders the attitude and skills required to cope with the highly complex and ever-changing environment. This type of on-the-job training is not provided to the managers at junior levels.
- **Leverage Commander's Intent**
  According to the military leaders, planning is everything. Followers in the military are clear about the intent of their leaders—their purpose, key outcomes, and the desired results. This clarity gives the followers an opportunity to adapt, develop, and succeed in accomplishing their own and organizational goals.
- **Organization of Tasks**
  The number of teams and missions is not equal in the military. Specific teams are created for achieving specific outcomes. The standards are universalized, and employees are trained in

---

[15] Webb, G. (November 14, 2011). 5 Things you could learn from military leaders. Retrieved from http://geoffreywebb.com//2011/11/14/5-things-you-could-learn-from-military-leaders/

a way that they are able to operate globally in multifunctional teams.
- **Use of Operators as Trainers**
  In the military, imparting training is the responsibility of the operators, and not that of the human resource (HR) department, unlike the situation in the corporate world. Operators have a better understanding of the skills required by the individuals, and, hence, they become the optimal choice for imparting training to the employees. The role of HR in the military is concerned with maintaining and tracking the progress record of the military employees.
- **Mission First Then People**
  The military organization has been established to serve people. The mission of the military leaders is of foremost importance, and even supersedes their personal interests.

ROLE OF MILITARY IN GENERATING GREAT LEADERS

The precepts and practices of military leadership are almost common all the world over, be it the American, or the Indian, or the Pakistani Army. They all promote the same values, such as *lead by example, know your job, value team spirit, imperative of loyalty to the organization, importance of moral and physical courage, and the capacity to take decisions.*[16] Also, the military leaders are given training on mission planning and execution of the given resources. The principles of military leadership are also applicable to corporate organizations, which imply that good military leaders will also become good corporate leaders.[17]

Military leaders are given training, which is enduring, carefully planned, has long tenures, and are costly. Military leaders are trained and developed in such a manner that they can handle stressful and

[16] Bennet, A., Bennet, D., & Lee, S. L. (2010). Exploring the military contribution to KBD through leadership and values. *Journal of Knowledge Management, 14*(2), 314–330.

[17] Vinay, S. (2010). Tenets of military leadership. *The Economic Times.* Retrieved from http://articles.economictimes.indiatimes.com/2010-12-07/news/28442818_1_corporate-leaders-ceos-tenets

unpredictable situations. Such an experience gives them the ability to handle stress, work with people as a team, develop strategies, and deal with the unknown.

- Military leaders excel in areas such as self-discipline, information gathering, situation assessment, communication, and intuition. They are continuous learners because as they shift higher in ranks, they are delegated more responsibilities, which enable them to embrace more leadership skills. The skills learnt during their military experience also help them after they retire from the military.
- The military plays a pivotal role in inculcating values in their employees, which form the asset of the military personnel. Military values are the guiding principles affecting the course-of-action of the personnel and the way one perceives the world around.
- In the military, the employees are given the freedom to apply leadership skills even in the early phase of one's career, which is not present in any other profession. Even the individuals at a lower level in the military have a high level of responsibility and authority.[18]
- Military employees continuously work in stressful and exhaustive situations, which enable them to develop skills required to perform paramilitary jobs, such as working in police departments. Employees from the military are preferred in police departments because they have the experience to work with weapons, along with the discipline and structure of military life, which is a prerequisite in the police department positions.[19]
- The challenges involved in military services require the rotation of leaders on different positions every two to three years. This rotation gives them a wide range of experience and helps

---

[18] Duffy, Tim (2006). *Military experience and CEOs—Is there a link?* Korn/Ferry International Report, Los Angeles, California.

[19] Ivie, D., & Garland, B. (2011). Stress and burnout in policing: Does military experience matter? *An International Journal of Police Strategies & Management, 34*(1), 49–66.

them in performing better than others, when they move to the corporate sector.[20]
- Both military and civilian leaders are faced with dangerous situations, though the nature of these situations is different in both the organizations. In spite of such differences, they resort to the same type of behavior. Military leaders apply five strategies to influence others and resolve conflicts, which can be learnt by the business leaders to prepare themselves for negotiations and bargaining processes. These strategies are maintaining a *big picture perspective, uncovering hidden agendas to improve collaboration, analyzing facts to get buy-ins, building trust, and focusing on process and outcomes*.[21]

LEADERSHIP EFFECTIVENESS

There have been several contradictory approaches to determine leadership effectiveness. The first approach to assess the effectiveness is to determine the relationship between leadership and intelligence quotient (IQ). However, the results of this approach have been very contradictory. The second approach to measure the effectiveness is concerned with the test of personality. However, this approach also had limited success. The third approach attempts to measure the leadership effectiveness through the experience and knowledge of the leader.

These approaches have limited success because they are based on academic tasks, rather than on practical tasks. Academic tasks are well defined, circumscribed, not motivating, and not relevant to many people's lives. In contrast, practical tasks are ill-defined, motivating, open, more concrete, and more relevant to people's lives. They are rather more practical. The abilities and skills required to perform academic tasks are not the same as the skills required to perform practical tasks. Another reason for the limited success of the

---

[20] Hemingway, S. Z. (March 6, 2007). Civilian vs. military: Who leads better? *Federal Express*. Retrieved from http://integrator.hanscom.af.mil/2007/March/03082007/03082007-17.htm

[21] Weiss, J., Donigian, A., & Hughes, J. (2010). Extreme negotiations. *Harvard Business Review*. Retrieved from http://hbr.org/2010/11/extreme-negotiations/ar/1

aforementioned approaches is that the traditional methods tend to involve only the adaptation of the environment that is, by changing oneself to suit the environment. However, actual leadership should be concerned with modifying the environment and shaping it in order to accomplish the organizational goals.[22]

Leadership is more concerned with human relationship between a leader and followers, who work together to achieve a shared purpose or goals. Leadership, being more concerned with people, should be viewed as an art to be developed, rather than a process to be mastered. Although, there have been many developed theories and principles for effective leadership, they are only a foundation to be used as a guideline by an effective leader, and not as a substitute for it. In addition to the leadership theories, leadership education should incorporate real-life experiences and exercises, case studies, etc.

To develop military leadership and enhance its effectiveness to the maximum, it is required to impart more training to the military force in their basic profession of fighting and handling weapons, and small units through coordination between the services. Further, the fighter leaders at a junior level should be lesser supervised, but empowered with more responsibilities, and should be made capable enough to take autonomous decisions with greater speed in decision making in highly demanding and complex situations. The leadership attributes required differ at the junior and higher levels. Junior fighter leaders are responsible for fighting and leading, while the leadership at a higher level requires thinking and planning about the future defense needs of the country, and how the military power has to be directed and guided to meet those needs, and the extent of flexibility it offers to cope up with the changing needs of leadership over time.[23]

---

[22] Hedlund, J., Hovrath, J. A., Forsythe, G. B., Snook, S., Williams, W. M., Bullis, R. C., Dennis, M., & Sterberg, R. J. (1998). *Tacit knowledge in military leadership: Evidence of construct validity* (Technical Report No. 1080). Alexandria, Virginia, United States of America: United States Army Research Institute for the Behavioral and Social Sciences. Retrieved from http://www.dtic.mil/dtic/tr/fulltext/u2/a343446.pdf (accessed on August 5, 2014).

[23] Singh, J. (June 7, 2004). Leadership for the 21st century. *The Indian Express*. Retrieved from http://www.indianexpress.com/oldStory/48442/ (accessed on July 31, 2014).

There have been many examples of effective leaders, and one of those many shining faces is that of the first Indian Field Marshal, Sam Manekshaw. He was a visionary and had never learnt to give up. His heroics were evident in the five wars that lasted for four decades. Under his leadership, the Indian Army was able to gain victory over Pakistan, which led to the liberation of Bangladesh in December 1971, then called East Pakistan. He had the experience, the shrewdness, and the ruthlessness of a very successful and respectable leader; he became the hero of the War of 1971. Furthermore, despite the long list of deeds and heroics that made him a field marshal, he was a man with personality and humility.

MILITARY IN THE CORPORATE

There are many parallels between business and military leadership. Even some of the words in business reflect a military orientation such as *waging price wars, strategic operations*, and *business tactics*, etc.[24]

Military organizations have been in the leadership-development business much earlier than the corporate world. Leading a military organization requires knowledge in various areas of this information-rich society. Military organizations operate in a highly uncertain environment characterized by high risk. Therefore, the military leaders require various tools and techniques that would assist them in facing these uncertain situations. In lieu of this, the military personnel are trained in a way to ensure their 24/7 readiness and commitment to the organization. In a similar fashion, the business leaders should also try to develop an adaptive culture to survive and succeed, knowing the fact that they, too, are operating in an uncertain environment and face a continuous increase in new type of competitors.[25]

A professor of Harvard Business School, Bill George (2010), pointed out the leadership effectiveness of the military and stated

---

[24] Rietsema, K. (2011). Celebrating memorial day: Business and military leadership. [Blog comment]. Retrieved from http://thecaalblog.com/aviation-and-aviation-leadership/leadership/celebrating-memorial-day-business-and-military-leadership.html

[25] Useem, M. (2010). Four lessons in adaptive leadership. *Harvard Business Review, 88*(11), 87–90. Retrieved from http://hbr.org/2010/11/four-lessons-in-adaptive-leadership/ar/1 (accessed on July 7, 2014).

that having an actual leadership experience, especially in the crisis situations, prepares the military leaders in a better and concrete way than the corporate experience in staff, consultation, or analytical roles. One can imagine the leadership experience of the military leader from the fact that a military leader who is twenty-two to twenty-six years old, gets an opportunity to lead around 100–150 persons in a crisis. This exposes military leaders to several aspects of leadership skills, and trains them for all sorts of contingencies that may arise during the course of operation. This type of experience is not available with corporate organizations, thus giving an edge to the military leaders over the corporate leaders.[26]

FUTURE OF LEADERSHIP DEVELOPMENT

With the increasing globalization, the world economy has seen many transitions. This has also led to changes in the way the organizations now build new markets and relate to their stakeholders. The entire world has become one, and the concept of "virtual world" has also evolved along with the concept of "physical world." Thanks to the various technological developments in the form of Internet, video conferencing, chat room, desktops net meetings, and groupware systems, such as Lotus notes, which have enabled the people all over the world to communicate with each other, irrespective of the physical boundaries. These developments will require significant changes to be done in the leadership pattern in the organizations, and, at the same time, grow and remain productive in the "old world order". The rapid advancement in information technology in the recent years has posed a great challenge in front of the leaders to be responsive to their followers and be in touch with them 24/7.

One of the best examples, where accessibility to information has changed a leader's way to lead the followers is the US military. Its future operating environment will be characterized by enhanced speed and complexity, wider dispersion of units, and increasing reliance on fewer systems and people. This would require people

---

[26] George, B. (2010). Why junior military officers become great business leaders. [Blog comment]. Retrieved from http://www.billgeorge.org/page/why-junior-military-officers-become-great-business-leaders (accessed on July 31, 2014).

to rely on more implicit form of training, and more informal form of acquiring knowledge and its dissemination.[27] Thus, managing the information (acquisition and dissemination) has become one of the main functions of the military leaders. To keep up with the continuous changes taking place in the dynamic environment, the military command system has adopted an absolute change. It now requires the followers to interpret the commands on the basis of the "intent," rather than simply "following the officer's directives." This change has been introduced realizing the fact that the conditions in the dynamic military environment in which the leaders are leading, may change even before the orders are executed.

The same concept can be applied to the corporate world where the employees are operating in a dynamic environment, in which they need to be made aware of their leader's intent, and, at the same time, have a discretion to take independent decisions, whenever required at the time of contact with the customers. However, the decisions taken should be within the boundaries laid down by the leader.[28]

LEADERSHIP: THE SAM'S WAY

Field Marshall Sam Manekshaw was the Army chief of India during the Indo-Pakistan war of 1971. He led India to one of the most important victories by forcing Pakistan to an unconditional surrender resulting in the liberation of Bangladesh in December 1971.[29]

(Box continued)

---

[27] Hedlund, J., Hovrath, J. A., Forsythe, G. B., Snook, S., Williams, W. M., Bullis, R. C., Dennis, M., & Sterberg, R. J. (1998). *Tacit knowledge in military leadership: Evidence of construct validity* (Technical Report No. 1080). United States Army Research Institute for the Behavioral and Social Sciences.

[28] Murphy, S. E., & Riggio, R. E. (2008). *The future of leadership development.* Mahwah, NJ: Lawrence Erlbaum Associates Inc.

[29] Gokhale, N. Remembering Sam Manekshaw, India's greatest general, on his birth centenary. http://www.ndtv.com. Retrieved from http://www.ndtv.com/article/india/remembering-sam-manekshaw-india-s-greatest-general-on-his-birth-centenary-503729 (accessed on April 3, 2014).

(Box continued)

> Tensions between the political leadership of Bangladesh (erstwhile East Pakistan) and Pakistan (erstwhile West Pakistan) was mounting in the early months of April 1971 which led to a war between the two nations. Millions fled to India and the Indian leadership called for a cabinet meeting to discuss the possibilities of waging a war against Pakistan. Manekshaw was called into this high profile meeting and was commanded by the Indian Prime Minister to wage a war against Pakistan. Manekshaw was required to give a decision at the spur of the moment and he knew that his decision will have an inevitable impact in the histories of both the nations, India and Pakistan.[30]
>
> Even though Manekshaw was the junior most member in the meeting, he showed unperturbed courage to say "no" to the Prime Minister. He backed up his refusal with relevant data, took an extreme decision, and stood by it.[31]
>
> Manekshaw knew that engaging in a war with Pakistan will result in conflicts with India's other neighbors. He reviewed all options related to internal stress, economic impact, and future reverberations of the war and refused to go on war with Pakistan. He made this bold choice and did not budge from his decision. He even extended his resignation to the Prime Minister if his decision were considered inappropriate, however he remained unfazed about changing his stance. He assured the gathering that given some additional time, he would be able to prepare his troops and even win the war for India.
>
> Such was Manekshaw's conviction that the Prime Minister allowed him to prepare his troops for war. Manekshaw's indelible

(Box continued)

---

[30] Bajwa, M. S. Sam Manekshaw's enduring legacy. http://www.hindustantimes.com. Retrieved from http://www.hindustantimes.com/punjab/chandigarh/sam-manekshaw-s-enduring-legacy/article1-1224879.aspx (accessed on June 1, 2014).

[31] Sinha, Lt Gen S. K. The Making of a Field Marshal. http://www.indiandefencereview.com. Retrieved from http://www.indiandefencereview.com/spotlights/the-making-of-a-field-marshall/ (accessed on April 18, 2014).

(Box continued)

> courage in these moments of extreme stress and his ability to refute the orders allowed India to win one of the most historic wars in which almost 90,000 Pakistani military personnel surrendered resulting in the liberalization of Bangladesh.[32]
>
> An effective leader is the one who weighs all the available options and distinguishes right from wrong irrespective of what his or her superiors might think or want. A great leader is one who remains steadfast on his/her decisions and his/her beliefs are not shaken by the most complex situations. Manekshaw was the epitome of such courage who stood rock solid in the most testing times and helped India to recreate history by redefining the political, social, and economic boundaries of the nation. By putting his foot down, Manekshaw reflected a mental attitude of determination and decisiveness. Such attribute reflects his courage, determination, and steadfastness even when he was pushed to the limits of endurance.

---

[32] IDR News Network. SAM Manekshaw on Leadership and Discipline. http://www.indiandefencereview.com. Retrieved from http://www.indiandefencereview.com/spotlights/sam-manekshaw-on-leadership-and-discipline/ (accessed on April 3, 2014).

2
LESSONS OF LEADERSHIP

Men make history and not the other way around. In periods where there is no leadership, society stands still. Progress occurs when courageous, skillful leaders seize the opportunity to change things for the better.

—Harry S. Truman

Leadership is a not a mere term that distinguishes a supervisor, or a head of state, or a chief officer from its juniors, followers, or subordinates. The term "leadership" has an abstruse meaning. Leadership is the driving force that makes people work toward a goal not only for an organization or a cause, but also for themselves. Leadership does not have a single definition; rather it cannot have, as *one size does not fit all*. Every institution and organization has a different mission, requirement, and goal. However, we can identify the leadership traits and principles that drive and create a leader.

Elliott R. Peterson elaborated the leadership traits by pointing out the following traits: (1) judgment, (2) justice, (3) decisiveness, (4) integrity, (5) dependability, (6) tact, (7) endurance, (8) bearing, (9) unselfishness, (10) courage, (11) knowledge, (12) loyalty, and (13) enthusiasm.[1]

---

[1] Peterson, R. E. (March 2012). Improve employee leadership with ideas borrowed from the military. Retrieved from http://www.astd.org/Publications/Magazines/TD/TD-Archive/2012/03/Improve-Employee-Leadership-with-Ideas-Borrowed-from-the-Military (accessed on August 6, 2014).

Judgment in leadership is assessing the situation in advance and making the right decision at the right time. Being just is a prime quality of a leader. This would imply that the leader will make fair decisions without any prejudices or any external influence. A leader must also be a prompt and an efficient decision maker. He/she has to assess a situation in advance, understand the situation that may arise, and make contingencies for the same. Another important quality of a leader is the ability to integrate the team or unit members. Integrity keeps members together and motivates them to help each other to achieve their individual goals, as well as organizational goals. A leader has to be a source of dependence for the members of the unit. This bestows upon the leader the duty not only to do the right things, but also to build trust and belief in the followers that he/she can always be relied on in any circumstance. Being tactful is another quality that leaders must have. Being tactful is the key skill in managing people sensitively and considerately.

As Gandhi puts it, "I suppose leadership at one time meant muscles; but today it means getting along with people."[2] A leader in the military has to face high demands of all sorts all the time. One must have endurance to be able to withstand any hardship and hindrance. A good leader never thinks about oneself. He/she will always prioritize the mission or the goal he/she is leading over his/her own wishes and desires. He/she is unselfish and thinks about everyone else first, and is, also, outright courageous. This implies that he/she will always stand up for what is right, irrespective of who is opposing him/her, and will always fight for it. All the aforementioned traits come with knowledge that comes from experience and by learning through examples. He/she is always loyal to his/her work, is always enthusiastic and passionate about what he/she does, and influences others to follow the same. He/she is driven by passion, empathy, and confidence.

PRINCIPLES OF LEADERSHIP

Principles of leadership are basically guidelines explaining how a leader applies the aforementioned leadership traits. Elliott R. Peterson listed the following principles used by the US Army. First principle is to

[2] Turk, W. (2007). Manager or leader?. *Defense AT&L*, 36(4), 20.

know oneself and constantly seek self-improvement. Second principle is to know one's soldiers and look after their welfare. Third principle is to ensure that the task assigned and undertaken is understood, supervised, and accomplished.[3]

Corporate organizations can use the aforementioned principles by merely replacing the term "leader" with "chief executive officer" (CEO) or "manager," and the term "soldiers" with "employees." The aforementioned principles, when incorporated within an organization's mission, lead to an increase in the productivity.

LESSONS FROM THE MILITARY LEADER

There is a lot that the corporate CEOs can learn from the military leaders. This can be implied from the statistics that the average tenure of a CEO with a military background is 7.2 years; whereas, the tenure of a CEO without a military background is 4.6 years.[4] Let us try and see what makes the military leaders so successful.

- **Creating a Personal Link**

    A leader or a supervisor in the military knows the members of the team/unit inside out. He/she not only knows their names and backgrounds, but is also aware of their potential and how each one would react in a given situation. It is his/her duty to understand the drawbacks of the mentees, and motivate them to improve on them, and improve their performance. To enable this, a leader builds a link with all the members of his/her team. You can affirm his/her link with others from the firmness of his/her handshakes. A leader's handshake indicates everything he/she wants to say. As he/she is the first or the immediate point of contact for them, his/her personal link with the members enables him/her to understand the situation better, and, also, in choosing the right people for the right tasks.

[3] Peterson, R. E. (March, 2012). Improve employee leadership with ideas borrowed from the military. Retrieved from http://www.astd.org/Publications/Magazines/TD/TD-Archive/2012/03/Improve-Employee-Leadership-with-Ideas-Borrowed-from-the-Military (accessed on August 6, 2014).

[4] Duffy, T. (2006). *Military experience and CEOs: Is there a link?*. Korn/Ferry International.

A corporate leader must also develop a personal link with his/her subordinates in the firm. He/she must know and understand who he/she is leading. This will help to build the right team and take individual decisions. This will help him/her judge the right people for the right task, thus fulfilling one of his/her important responsibilities.

- **Making Right Decisions**

Making right decisions at the right time is the most important task of a leader. A leader has to constantly make both easy and difficult decisions. He/she has to make the decision to choose the right task and to appoint the right people to perform it. He/she has to choose the right strategy to derive the desired results. He/she has to strategize, act, and delegate. Moreover, he/she has to act fast before he/she misses the opportunity. A short delay or a mistake from the leader's front can lead to huge losses in the form of loss of opportunities.

- **Focusing on the Mission**

A leader is driven by the mission he/she is striving for. The mission lures him/her to do all the things he/she needs to do to achieve it. A leader has to incorporate all his/her leadership qualities and incorporate them with the mission of the organization he/she is leading to derive results meeting the organizational goals. Moreover, he/she has to follow an order while performing his/her duties. The order is that he/she first thinks about the mission of the organization, then he/she thinks of the welfare of the team or unit, and then finally he/she thinks of himself/herself. He/she is always his/her last priority because the goal is to achieve a goal that is bigger and more important than him/her.

- **Making the Right Strategies**

Making the right strategy involves using the acquired knowledge, experience, observations, and using the right strategies. A leader is responsible for choosing the right strategy to accomplish a given goal. He/she has to be proactive and has to understand the moves, strategies, and think of counter strategies used by the rivals/competitors. Using the right strategy is the determining factor for the success of a leader's decisions. It is also important that a leader conveys the strategic intent to others. This will

help and enable others to coordinate and act in a way that leads to achieving the common organizational goal.
- **Strengthening Communications: Feedback and Participation**
A leader has to strengthen his/her communication with the team members by encouraging them to actively participate in the decision-making process via meetings and discussions. This enables the leader to understand what his/her team members think. Also, to evaluate their opinions and views, he/she must ask them to fill feedback forms. Feedback forms help the leader in evaluating the team, understand the team dynamics, and enable him/her to mould his/her actions accordingly.
- **Being Present at the Right Place**
A leader has to be present at the right place at the right time. A general perception of a leader's position is always being at the front. However, a leader does not always have to be at the front. A leader does not have to be at the front of everything. Rather, he/she has to be at a place where friction arises; in other words, a place where the greatest difficulty could arise. This is basically positioning his/her existence on the basis of highest priority.

KEY FEATURES OF THE MILITARY AS AN ORGANIZATION[5]

There are certain key features of the military as an organization, which make it unique and different from other organizational forms. Wong, Bliese, and McGurk stated the following features of the military:

- **Diverse Collection of Members**
A military consists of people from diverse fields, such as army, navy, and air force, who have different cultural backgrounds and roles to play. Hence, each domain has its own unique definition of leadership.
- **Size of the Organization**
The size of the military is huge compared to other organizations, such as postal services, and law and educational

[5] Wong, L., Bliese, P., & McGurk, D. (2003). Military leadership: A context specific review. *The Leadership Quarterly, 14*(6), 657–692.

departments. In case of military organizations, leadership has a greater importance because leaders at both junior and senior levels have to command a large number of subordinates. The larger the size of an organization, the more complex becomes the role of a leader.

- **Organizational Forms**
  The military organization follows a traditional form of structure, where there is clarity in power delegation across all levels and a strict code of conduct is followed by all the members of the system, both during and after the work hours. This code of conduct and power structure is followed across every domain and hierarchal levels in the military.
- **Leadership, a Mainstay of the Military**
  Leadership evolved in the military even before researches were undertaken in the corporate or education field. Attempts have been made by the military to develop leaders through their formal educational process, operational assignments, and self-development. The US Army officials, for example, spend three years out of their twenty years of career in the army schools to learn and develop the required skills. This indicates that the military personnel are trained to be leaders for a longer period than in any other organization.

MILITARY PRODUCES GREAT LEADERS

The military has been an efficient creator of great leaders. A study suggests that CEOs with a military background tend to deliver stronger performance as leaders. Also, they tend to survive longer on the job, and skills learned from the military training enhance their success in the corporate world.[6] Kolidtz has attributed several reasons to this:[7]

- The military has a systematic, progressive, and carefully planned training, educational and experiential events. These

[6] Duffy, T. (2006). *Military experience and CEOs: Is there a link?*. Korn/Ferry International.
[7] Kolditz, C. T. (February 6, 2009). Why the military produces great leaders. [Web blog comment]. Retrieved from http://blogs.hbr.org/frontline-leadership/2009/02/why-the-military-produces-grea.html

training sessions are extended for longer periods and are costlier than any other training programs in any of the other government organizations or private institutions.
- The responsibility, power, and authority that the military personnel in the lower rank have are higher than the responsibilities that the employees with a similar rank in any other institution may have.
- The pillars on which the whole military organization stands are service, duty, and self-sacrifice, and an oath is taken to commit to those pillars. The concept of leadership in the military is not only restricted to the employees' professional lives, but is also extended to their families. In the time of crisis, leadership as an operational function influences the physical well-being and survival of the leader as well as the follower. Transactional form of motivation, such as monetary benefits or punishments, is of no influence in this situation. Leaders need to acquire the trust and confidence of the subordinates and transform them into religious followers. Hence, the transformational style of leadership is more prominent in the military.

Military service provides its employees with opportunities to learn and apply leadership skills at a very early stage of their career. This is not so in any other profession or organization. The effectiveness of military leadership can be viewed from the fact that the likelihood of a military officer (among US male officers) becoming the CEO in some other company is higher than that of other members belonging to other professions. The core traits of the military officers, such as discipline, ambition, and goal-oriented passion, give them an edge over others. The leadership skills and experiences the military officers have learnt during their tenure in the military help them in climbing the corporate ladder with greater confidence and ability.[8]

---

[8]Duffy, T. (2006). *Military experience and CEOs: Is there a link?*. Korn/Ferry International.

First Indian Field Marshal, Sam Manekshaw: Narration of a Magnificent Experience

The first Indian Field Marshal, Sam Manekshaw, recollected his experience of the 1971 War in a lecture he delivered at the Defense Services College, Wellington, in Tamil Nadu, on leadership and discipline in 1998. He was an able speaker. He did not ask for attention; he made the people listen to him because each word had the experience, the shrewdness, and the ruthlessness of a very successful and respectable leader, the hero of the War of 1971. He had mannerisms of a gentleman and used humorous one-liners like an entertainer. He deserved respect for whatever he spoke: his army life, the war conditions, and the conversation with the prime minister. The crowd was all ears for him as he motivated them, and stirred up their thinking and ambition. He had led an army of thousands and thousands were present to listen to him. A few excerpts from his speech would be worth mentioning. He addressed the youth about how there had been a dearth of leadership and how it could be overcome.

In the words of Sam Manekshaw (direct quotes from the website Indiandefencereview.com):

> The problem with us is the lack of leadership.
> When I say lack of political leadership, I do not mean just political leadership. Of course, there is lack of leadership, but also there is lack of leadership in every walk of life, whether it is political, administrative, in our educational institutions, or whether it is our sports organizations. Wherever you look, there is lack of leadership. Leaders are not born, so can leaders be made? My answer is yes. Give me a man or a woman with a common sense and decency, and I can make a leader out of him or her.
> What are the attributes of leadership?
> The first, the primary, indeed the cardinal attribute of leadership is professional knowledge and professional competence. Now, you will agree with me that you cannot be born with professional knowledge and professional competence, even if you are a child of Prime Minister, or the son of an industrialist, or the progeny of a Field Marshal. Professional knowledge and professional competence have to be acquired by hard work and by constant study. In this fast-moving technologically developing world, you can never acquire

sufficient professional knowledge. Professional knowledge and professional competence are a sine qua non of leadership. Unless you know what you are talking about, unless you understand your profession, you can never be a leader. Every time you go around somewhere, you see one of our leaders walking around, roads being blocked, and transport being provided for them. Those, ladies and gentlemen, are not leaders. They are just men and women going about disguised as leaders.

What is the next thing you need for leadership?

It is the ability to make up your mind to make a decision and accept full responsibility for that decision. Have you ever wondered why people do not make a decision? The answer is quite simple. It is because they lack professional competence, or they are worried that their decision may be wrong and they will have to carry the can. According to the law of averages, if you take ten decisions, five ought to be right. If you have professional knowledge and professional competence, nine will be right, and the one that might not be correct will probably be put right by a subordinate officer or a colleague. But if you do not take a decision, you are doing something wrong. *An act of omission is much worse than an act of commission. An act of commission can be put right. An act of omission cannot.* Make a decision and having made it, accept full responsibility for it. Do not pass it on to a colleague or subordinate.

So, what comes next for leadership?

Absolute honesty, fairness and justice—we are dealing with people. Those of us who have had the good fortune of commanding hundreds and thousands of men know this. No man likes to be punished, and yet a man will accept punishment stoically if he knows that the punishment meted out to him will be identical to the punishment meted out to another person who has some Godfather somewhere.

We, in India, have tremendous pressures—pressures from the government, pressures from superior officers, and we lack the courage to withstand those pressures. That takes me to the next attribute of Leadership—Moral and Physical Courage. I will lay emphasis on moral courage.

What is moral courage?

Moral courage is the ability to distinguish right from wrong and having done so, say so when asked, irrespective of what your superiors might think or what your colleagues or your subordinates might want. A "yes man" is a dangerous man. He may rise very high; he might even become the Managing Director of a company. He may do anything but he can never make a leader because he will be used by his superiors, disliked by his colleagues and despised by his subordinates.

I am going to illustrate, from my own life, an example of moral courage. In 1971, when Pakistan clamped down on its province, East Pakistan, hundreds and thousands of refugees started pouring into India. The Prime Minister, Mrs. Gandhi had a cabinet meeting at ten o'clock in the morning. The following attended: the Foreign Minister, SardarSwaran Singh, the Defense Minister, Mr. Jagjivan Ram, the Agriculture Minister, Mr. Fakhruddin Ali Ahmed, the Finance Minister, Mr. Yashwant Rao, and I was also ordered to be present. There is a very thin line between becoming a Field Marshal and being dismissed. A very angry Prime Minister read out messages from Chief Ministers of West Bengal, Assam and Tripura, that hundreds of thousands of refugees had poured into their states and they did not know what to do.

The prime minister asked him to enter Pakistan, irrespective of whether it would lead to a war or not. The field marshal, completely aware of his troops, their capabilities, and the conditions that they were going to face, denied the possibility of fighting and winning the war. He remained calm and showed that attacking at that point of time would mean a hundred percent defeat, being weak on the China front—in defense and in supplies. The prime minister asked whether that was the truth, to which Manekshaw replied:

> Yes, it is my job to tell you the truth. It is my job to fight and win, not to lose.
> Let me guarantee you this that if you leave me alone, allow me to plan, make my arrangements, and fix a date, I guarantee you a hundred percent victory.

He portrayed a magnificent example of moral courage. Another point that he adds to the leadership skills is the physical courage.

> It is one thing to be frightened. It is quite another to show fear. If you once show fear in front of your men, you will never be able

to command. It is when your teeth are chattering, your knees are knocking, and you are about to make your own geography—that is when the true leader comes out!

Loyalty is the next very important trait that a leader must have. A leader, if expects loyalty from his/her subordinates, should also be loyal to them in case they need it. Moreover, a leader should be an example to his/her subordinates for whatever he/she does. Any action taken which is in contrast to what they actually are is derogatory to his/her authority and image as an established leader. He/she should talk and act precisely to the point that he/she is supposed to. But his/her mind should always be on the future, planning ahead, making strategies. And all of this should be followed up with proper discipline. He ended the speech with a talk on discipline.

He says:

> What is discipline? Please, when I talk of discipline, do not think of military discipline. That is quite different. Discipline can be defined as conduct and behavior for living decently with one another in society. Who lays down the code of conduct for that? Not the Prime Minister, not the Cabinet, nor superior officers. It is enshrined in our holy books; it is in the Bible, the Torah and in the Vedas, it is in the teachings of Nanak and Mohammad. It has come down to us from time immemorial, from father to son, from mother to child. Nowhere is it laid down, except in the Armed Forces, that lack of punctuality is conduct prejudicial to discipline and decent living. (This has been directly quoted from the article "Sam Manekshaw on Leadership" from the website Indiandefencereview.com.)[9]

Lieutenant General Russel Honoré (retired), the former commander of Joint Task Force—Katrina, who oversaw military relief efforts after Hurricanes Katrina and Rita, had a question and answer (Q&A) session on leadership lessons.

He stresses over the fact that it is not easy for an army general to give orders, and, hence, assume that the work is done. People

[9] Indian Defence Review. (April 3, 2014). *India Defence Review*. Retrieved from India Defence Review Website: http://www.indiandefencereview.com/spotlights/sam-manekshaw-on-leadership-and-discipline/ (accessed on August 7, 2014).

in uniform are no different from others. A given order requires the supervision of a leader to be executed perfectly. He stresses that above all, one should not be a boss to people. A leader's job is strategic: to set people on the right path and "to do the planning and then to motivate the execution."

According to him, leadership means forming a team and working toward common objectives that are tied to time, metrics, and resources. The purpose of the commander and the staff is to do the planning and then to motivate the execution. Many times, leaders say about the plan of the work that they are going to do, but it fails if the execution is not tracked.[10]

CORE MILITARY LEADERSHIP COMPETENCIES

Core leadership competencies of the military personnel are the driving factors which keep them going. All these competencies are idealistic and attainable. These competencies can give an idea to the upcoming business managers and CEOs about what makes the military leaders successful and learn from them.

- **Energy Level and Endurance**
  When it comes to the energy levels and endurance of a leader, there can be no better example than that of a leader in the military. A leader or a commander in the military is the most energetic and inspirational person one may come across. The credit for this goes to the military training and the experiences the person has had. Not only is he physically fit, but he is also perpetually enthusiastic and self-motivated. There are several factors that contribute to this. One of them could be that a leader in the military has strived for his/her passion and has followed it throughout. He/she is driven by the mission, goals, and the values of the military, and strives to give his/her best and achieve it.

---

[10] Anonymous. (2009). A military general's leadership lessons. Retrieved on June 26, 2013, from http://businessjournal.gallup.com/content/113629/military-generals-leadership-lessons.aspx#1

- **Confidence**
Confidence is the first thing you will notice in a military leader. This will be reflected in his/her first point of contact, such as, while addressing, greeting, and making firm handshakes. A military leader can see through the other person without having any reflection on his/her face and gestures. A leader in the military is trained to be confident. Also, he/she has witnessed and experienced several difficulties in his/her lifetime as a soldier, man/woman of the family, and as a team member in the military. His/her confidence is a reflection of his/her capabilities and the lessons he/she has learnt in life. The confidence of a leader has the power to motivate, influence, and guide several followers and onlookers.
- **Emotional Stability**
A military leader undergoes several emotions from the beginning of his/her entry in the military. A military leader first undergoes a strict disciplined routine, along with the strenuous physical activities. Besides, he/she also faces harsh punishments for not following the minutest orderliness. Simultaneously, he/she also has to keep away from the family for long durations, which has been their emotional support system throughout. As they move up in their careers, they get used to the hardships they face individually, but soon they are sent to war fields, where they are exposed to brutal and traumatic incidents. Long after a soldier in the military has survived all the aforementioned emotional turmoil, he/she moves a step ahead toward becoming a military leader. A military leader is emotionally more stable, as he/she has been trained to do so by winning over and learning to control even his/her most intense emotions.
- **Motivation**
Military leaders are highly motivated and are highly motivating. Their speeches and commands always prompt you to move forward in life without holding back and looking back. They always command the juniors to march forward and accomplish the aimed tasks and goals. A military leader has to be highly motivating, as many officers who are not able

to deal with the pressure in the military tend to succumb to them and surrender. It is the leader's role and duty to motivate such officers, and get them back on track and in the team. A leader is able to motivate others because he/she is a living example of someone who has successfully managed sailing through the tough journey in the military. As a motivator, he/she always motivates the juniors to keep going through the difficulties to attain victory.

- **Building Integrity**
  A leader in the military manages a team, or a unit, or a division. The tasks assigned to the officers at the army level are very crucial and sensitive. Accomplishing tasks is a group effort and not an individual task. It is essential that he/she builds the feeling of integrity among the team members, so that they help, guide, and motivate each other in times of need. If the team is not integrated, it will be easy for the opponent party to attack their unit and defeat them.
- **Strong Communication**
  A leader in the military has to be an effective communicator. The effectiveness of his/her communication skills can be seen from the effortless clarity and sternness in his/her voice. He/she speaks and commands in a way, which leaves no scope for ambiguity and doubt in the minds of the listeners. As a leader, he/she makes sure that once an order is given, it is obeyed. Developing strong communication skills is also a part of his/her training in the army. In the army, the leader has to give commands throughout. He/she has to be sure of whatever orders he/she gives. The leader is often involved in making strategic plans. He/she has to make sure that he/she communicates all the important details to everyone, and, also, that the officers have understood them and are following it.
- **Working for Welfare of Subordinates**
  A leader in the military, after prioritizing and acting in favor of the military, has to act in favor of the welfare of the subordinates. While making or being a part of any decision, he/she has to be sure that a decision or an amendment is first in

favor of the military's mission, and then he/she has to also ensure that he/she is simultaneously taking care of his/her subordinates' welfare. He/she has to be harsh and strict, but, at the same time, he/she also has to be sure that the officers are taken care of.

- **Lead by Example**
  The most efficient way in which a leader can lead is by being an example to others. A leader knows that he/she has an effect on others. His/her temperament—optimism or pessimism—spreads like a virus among the followers. A leader is always aware that he/she is accountable for his/her actions, and, even, thoughts. A leader has to practice discipline and sacrifice happily to be able to influence others. He/she has to show others that doing the right thing is always the right thing to do to go ahead in life. Being a leader comes with a great responsibility. One has to understand that one must obey one's own principles and values throughout one's life. Leading by example is the mantra of being a leader for life.

- **Create a Positive Work Climate**
  A good leader always creates a positive work environment. In an army, when a supervisor walks in, all the officers straighten up. He/she can be strict, and still be very effective and positive. A positive work climate is a climate that is conducive to increasing performance of the teams. Officers are scared of their leader, but, simultaneously, they also respect him/her. They are aware of the fact that their leader can see through them. A leader also creates a positive work environment by making the environment friendly. This can be done by having frequent interactions with the employees and by making them feel important. The leader of a group or a unit acts as a guide for his/her subordinates. He/she counsels them in their professional and personal matters too. This becomes his/her key role as most of his/her subordinates have left their homes, and need emotional support and backing. A leader, here, plays the role of a guide and helps his/her followers by motivating them during rough times.

LESSONS OF LEADERSHIP

- **Creating Leaders**
  Creating leaders is another duty of a leader. A leader, after leading for many years, has acquired a unique set of knowledge, principles, guidelines, etc. This is based on the crux of whatever he/she has learnt in his/her entire lifetime. It is the duty of a leader to ingrain whatever he/she has learnt in suitable prospects and get them ready to lead when he/she is close to retirement. A leader has the power to create more leaders and he/she must do so. Leaders have to select the right people and then train them to make leaders for tomorrow.

LEVELS OF LEADERSHIP

As per Army Leadership (AR 600-100), there are different leadership levels, which differ in their requirements in terms of combination, length, breadth, and scope. As a leader moves up the level, the complexity and his/her independence in the assignments also increase. This requires new and more skills, authority, and accountability. This progress happens in three stages or levels, which are:

- **Direct Level Leadership**
  Direct level leadership is a frontline level leadership, which includes leaders leading groups as well as divisions. Direct level leaders are responsible for building teams that are cohesive, empowering their subordinates, and developing right strategies to accomplish the organizational goals. This level of leadership requires direct face-to-face interaction between the leader and the subordinates. Leadership at this level influences human behavior, beliefs, and attitudes of other members. However, the planning and accomplishment of objectives done at this level are of short term, which range from a period of three months to one year.
- **Organizational Level Leadership**
  Organizational level leadership is normally required in a more complex organization. This includes leaders leading teams at corporate levels and at the directorate level. In accession to

the roles of the direct level leader, organizational level leaders need to arrange for the resources of the organization, manage multiple priorities, establish long-term goals, and empower others in accomplishing them. They need to pay more attention to planning, making policies, integrating systems, etc., and not in interpersonal interaction with the subordinates. Their main focus is toward mid-range planning, which ranges from one year to five years or more.

- **Strategic Level Leadership**
Strategic level leadership is the highest level that a leader can attain. This includes military and corporate leaders at division and corporate levels, and national level. Strategic leaders are responsible for setting up the organizational structure, allocating resources, and articulating the strategic vision. They assess the external environment, such as market and political conditions, and try to understand its impact on the organization, and, accordingly, alter their plans and strategies. These leaders have associations with people of high-level leadership and other dignitaries. Decisions made by these leaders have a long-term political, organizational, and personal effect, as they have a strong followers' base. Leaders in this level are competent enough to solve complex problems and make strategic decisions, using their analytical abilities and critical reasoning. Their goals and plans are long term, and are also responsible for developing a long-term vision for the organization ranging from five years to twenty years or more. These leaders teach, train, and motivate others by setting their own examples.[11]

HOW CAN A LEADER IMPROVE?

Leaders in any organization will need to continuously improve their skills and enhance their knowledge to enable them climb the ladder from the direct level leadership to strategic level leadership.

[11] Army Regulation (AR 600-100). (March 8, 2007). Headquarters, Department of the Army, Washington, DC. Retrieved from http://fas.org/irp/doddir/army/ar600-100.pdf (accessed on August 5, 2014).

Following are a few aspects laid down by Gary Yukl that can help a leader in improving himself/herself:[12]

- **Self-Awareness**
  Self-awareness is about knowing oneself better. It is about assessing and understanding one's weaknesses and strengths, and using this knowledge to improve one's performance. It helps the leader understand his/her emotions and complexities, helping him/her take better decisions without having any prejudice. If one knows how one would react to a particular situation, he/she can choose to take the lead, or even choose to back out if he/she thinks someone else can lead better. Understanding oneself helps the leader in understanding others better. Self-awareness is the first step toward self-development.
- **Developing Relevant Skills**
  Leaders have to constantly learn and upgrade the required skills. As technological developments and socioeconomic changes occur, leaders must update themselves with skills that are compatible with the changes. They should identify the skills that suit them the best and learn them through workshops or self-initiatives.
- **Don't Let Your Strength Become Your Weakness**
  This means that one has to change and adapt to changes. If a strategy once used proved to be successful, it is not necessary that it will work every time. Leaders need to be flexible and not be stuck with just one skill or strategy they have. As the leaders move upward in the career graph, they will have to face diverse problems from diverse domains. Thus, leaders will have to use different skills and strategies for managing different problems and issues.
- **Compensate for Weakness**
  We all have weak points. A leader will always think of how he/she can make up for the weakness he/she has. A leader can overcome his/her weakness, but this will take some amount of

---

[12] Yukl, G. (2010). *Leadership in organizations*. Essex, England: Pearson Education, Inc., Prentice-Hall.

time. In the meanwhile, he/she should handover the responsibilities in which he/she thinks he/she cannot perform well to someone else, a subordinate, or another manager, who is good at it. A leader looks at accomplishing the task efficiently, irrespective of who does it.

LEADERSHIP SKILLS

Leadership skills have many sub-skills that a leader needs to learn to become an efficient, influential, and a successful leader.

- **Technical Skills**
  Leaders in the military are technically well endowed because of their exposure to high-level tasks, which require a higher degree of skills, ability, and training. Technical skills refer to knowledge about equipments required for specialized activities, organizational methods, processes, employee characteristics, management systems, etc. These skills can be acquired by obtaining formal training, education, and experience. Leaders need to possess extensive technical skills, as they will need to train, guide, and supervise the tasks assigned to people. Technical skills also help in evaluating and monitoring others' performance.
  In case of business firms, leaders/managers are required to have a sound knowledge of the new products or services launched in the market. They should possess the technical skills that enable them to conduct right research, and then make strategic plan to establish itself or a new product in the market.
- **Conceptual Skills**
  Conceptual skills are also known as cognitive skills. These skills involve good judgment, foresight, creativity, etc. Higher cognitive complexity enables the leader to gauge various dimensions and effects of a change or a new implementation. This skill is important for effective planning, organizing, and problem solving. This skill is also very essential for strategic planning, as it helps the leader in foreseeing issues, anticipating changes, and indentifying opportunities.

- **Interpersonal Skills**
  Interpersonal skills are about analyzing and understanding human and group/team behavior. It helps the leader in understanding attitudes, feelings, motives, etc. of individuals. This also brings the ability to communicate efficiently and influentially. A leader with this skill is capable of building strong and cooperative relations with people at different levels. Interpersonal skill also includes self-monitoring. This enables the leader understand his/her behavior and helps him mould it as per the need of the situation. Leaders with this skill can play a major role in building employee-related policies and behavioral standards.
- **Emotional Intelligence**
  It is after 1995, when Goleman published his book on emotional intelligence (EI), that the importance of being emotionally intelligent even in organizations was realized. Emotions are an important part of everyone's life. Their existence and importance cannot be ignored in the organizational space. A good leader is emotionally intelligent, which means that he/she is in good control of his/her feelings. A good leader uses his/her emotional intelligence for the welfare of the followers. EI enhances the cognitive skills of the leader. EI encompasses the qualities of self-awareness, self-regulation, and communication skills. It can have a direct impact on the task performance and interpersonal relationships of the leader. EI makes the leader empathetic, which enables him to understand others better and listen to others carefully. It also enables the leader to build enthusiasm, optimism, and motivation in the team.
- **Social Intelligence**
  Social intelligence is driven by two main components—social perceptiveness and behavioral flexibility. Social perceptiveness has to do with understanding the needs, problems, and opportunities related to a group or the organization. Social intelligence is also about understanding group member characteristics, social relationships, and formulating appropriate responses. It requires the knowledge of group processes, organizational culture, and structure. The second component,

behavioral flexibility, is about the ability of the leader to adapt and change according to situations. High level of behavioral flexibility enables to understand and apply various dimensions of leadership as per the need of the hour.

- **Constant Learning**
  An efficient leader knows that there is never an end to learning. He/she tries to learn from every aspect of his/her life. He/she learns from his/her and others' experiences. He/she is also always keen to learn new techniques, strategies, etc. He/she is driven by the constant feeling of improving himself/herself, and is not hesitant in learning from anyone, even from his/her subordinates. This ability enables them to adapt to new situations, constantly innovate, and be flexible and open to new ideas and directions. A leader constantly learns because he/she constantly assesses his/her performance and is looking for ways to improve them. He/she is self-analytical and self-critical. To adapt and accommodate to the changing situations and conditions, it is important that a leader is up for learning anything and everything that helps him/her achieve his/her and the organization's goal.

BASIC LEADERSHIP STYLES

There are four widely known leadership styles. Each of them is explained as follows:

- **Autocratic Leadership Style**
  Autocratic leadership style is a classical approach of leadership. In this style, the leaders or managers retain majority of the power of decision making. Followers or subordinates have no say in making changes or amendments in the organizational processes. In this case, the leader does not believe in consulting others, and they are also not allowed to give inputs. Subordinates are merely expected to perform their tasks as per the instructions given, without raising a question or a doubt. This style uses a set of penalties and remunerations that are allocated as a token of mistakes and good performance respectively.

This style of leadership has been highly criticized for its unemphatic approach. Leaders following this style do not trust their subordinates, and believe in treating them with punishments to influence and push them to perform better. Today, this style is not used in organizations. This style of leadership is not recommended for any leader.

- **Bureaucratic Leadership Style**
  A leader who follows the bureaucratic style strictly adheres to the rules, guidelines, procedures, and policies. A leader is the one who brings change by influencing others for the betterment of the members and the organization. A leader must follow the rules and policies of the organization, but there are unique situations wherein the given rules and policies cannot be controlled by following them. This is where the leader plays his/her role of doing the right thing for the betterment of the organization, even though a provision has not been made for it in the guidelines.

- **Laissez-Faire Leadership Style**
  Laissez-faire is an economic doctrine which believes that there should be a minimum or no interference from the government. In case of the corporate firms, this would mean that the authorities of the organization do not control or monitor the employees and give them the freedom to perform without any given direction or guideline. Here, the decision-making power rests in the hands of the employees, and they are given the liberty to take initiatives, and make decisions to resolve issues.
  This style of leadership is not favorable to all types of organizations. This style can be applied where the employees are highly skilled, trained, and motivated. It can be applied to an organization, such as an advertising agency, that involves high level of creativity on the part of the employees.

- **Democratic Leadership Style**
  The democratic style of leadership is a style where a leader encourages participation of the employees by asking for/giving feedbacks and opinions. It gives some amount of power and authority to the employees in the decision-making process. In this style, the leader effectively and efficiently

communicates with the members of the organization, and keeps them updated about the new developments. In the democratic style, the leader is concerned with what the others have to say and think. He/she believes in them and is successful at encouraging the employees to perform better.[13]

The aforementioned leadership styles are the traditional ones. All these styles cannot prove to be successful if followed in entirety. Bits and pieces of each of these styles may be adopted, depending on the nature and the mission of the organization. However, with an increase in diversities across the globe, new styles of leadership have emerged, which can fit in different organizations with different and higher dynamics.

The role the military has played and is playing in the corporate sector is more than what we can imagine. The proportion of CEOs who have served previously as military leaders is much higher than that of the CEOs with a civilian background.[14] This clearly implies that the military leaders are good not only at leading the military, but also at leading the business firms. The leadership skills and the experiences that a military leader has increase his/her ability to climb the ladder in the corporate world after leaving the military.

CEOs with civilian backgrounds or prospective civilian leaders cannot gain the military experience in the exact way military leaders do. However, the least one can do is learn from their lives, experiences, and stories. There are many lessons that one can learn from the mistakes and risks taken by the military leaders.

"There are some who've forgotten why we have a military. It's not to promote war; it's to be prepared for peace."[15]—President Ronald Reagan.

---

[13] Gastil, J. (August 1994). A meta-analytic review of the productivity and satisfaction of democratic and autocratic leadership. *Small Group Research, 25*(3), 384–410.

[14] Duffy, T. (2006). *Military experience and CEOs: Is there a link?*. Korn/Ferry International.

[15] Reagan, M. (2004). In the words of Ronald Reagan: The wit, wisdom, and eternal optimism of America's 40th President. Nashville, Tennessee: Thomas Nelson Inc.

LEADERSHIP IN MILITARY, DEVELOPMENT, AND TRAINING

Leader Development

Leader development is a continuous, steady, and progressive process based on army values that develops soldiers and civilians into competent and confident leaders, capable of decisive action. Leader development is achieved through permanent integration of the knowledge, skills, and experiences gained through institutional training and education, organizational training, operational experience, and self-development. Commanders and other organizational leaders are responsible for producing competent, confident, and agile leaders, who are bold and quick in dynamic and complex situations.

Army Training and Leader Development Model

The army training and leader development model labels necessary areas of training for soldiers and developing leaders. The model projects lifelong learning and identifies three developmental domains that shape critical learning experiences: operational, institutional, and self-development. The model demands the development by competent and confident leaders, and depicts a continuous cycle of education, assessment, and feedback. For each domain, a specific measurable action is required, and each domain should use assessment and feedback from various sources to maximize mission readiness and to develop leaders.

- **Training and Leader Development Domains**
  According to a study conducted by the US army in 2009, the three focus areas of leader development (institutional training and education, operational assignments, and self-development) are productive, forceful, and interconnected. At the institutional level, individual gains knowledge and skills and then enhances them by practicing during operational assignments. Self-development enhances, sustains, and expands the knowledge, skills, and abilities gained from assignments and institutional learning.

  (a) **Institutional Training and Education**
  The army's school system enriches a leader with the education (how to think) and training (how to do) needed to perform

duty and satisfy the position's requirements. Leaders attend institutional training courses following appropriate career development models.

(b) **Operational Assignments**
In operational assignments, institutional training and education are put into practice by placing leaders in situations that require them to apply the knowledge and skills they have acquired. Pairing of repetitive performance with self-awareness, assessment, and feedback not only enhances leadership skills, but also broadens knowledge, shapes attitudes, and subsequent behavior.

(c) **Self-Development**
Self-development mainly focuses on maximizing leadership traits, reducing weaknesses, and achieving individual leadership development goals. Self-development is a never-ending process that begins during institutional training and education, and continues until operational assignments. Self-development also prepares an army leader for future responsibilities.

- The main reason for setting up of army training and leader development management process was to improve training and leader development policy, strategy, and capability needed to provide trained and ready soldiers, leaders, and units to combatant commanders. The management process has councils of colonels (COC) on the top, and it provides recommendations to army leadership through the Training and Leader Development General Officer Steering Committee.

  It is the responsibility of all the leaders to guide and mentor their juniors, and develop them to the fullest extent possible. Counseling and coaching are the two important tools, which a leader can use to facilitate development. Counseling can be used as a standardized tool to provide feedback to a subordinate, and coaching can help a junior in functioning through a set of tasks.[16]

---

[16] Anonymous. (2009). Army Training and Leader Development. Retrieved from http://www.apd.army.mil/pdffiles/r350_1.pdf (accessed on June 27, 2013).

ETHICAL CONDUCT: LEADING BY EXAMPLE

Major Vikram Batra PVC (September 9, 1974—July 7, 1999) was an officer of the Indian Army, who was posthumously awarded with the Param Vir Chakra for his actions during the 1999 Kargil War in Kashmir between India and Pakistan.

Fortnight after he became the face of the Indian soldier in the Kargil war, Vikram Batra died. On July 7 night Vikram Batra and his team recaptured Peak 4875 and July 8 morning after been severely injured he died. Vikram with his men had begun a tortuous climb to strengthen the flanks of the Indian troops fighting the Pakistani troops. When Vikram and his fellow officer, Anuj Nayyar, were attacked, they retaliated aggressively. They forced Pakistani troops to retreat after an intense exchange of fire. They were also successful in clearing the enemy bunkers. In a fateful accident when a junior officer injured his legs in an explosion, Vikram attempted to rescue him out of the danger. Though the junior officer suggested him to leave, Vikram following his principles persisted and saved the young Lieutenant. However in the process, a bullet pierced through his chest. Though India won back Peak 4875, it lost a brave son of India in the form of Vikram Batra.[17]

Such ethical decision taken by him required diligent effort and courage. The strong instinct to survive urges us to avoid situations which may put our life into risk. Standing firm on one's principles is not a God gift. It grows along with the individual over years of training. Facing corporal danger and being ready to sacrifice life is something more than just performing the task and that is what differentiates warriors from mortals. The staunch commitment to duty can be performed only by an individual who has a well-defined tenet of ethics. From the above incident that happened in Vikram' s life he could have

(Box continued)

---

[17] Batra, G. L. Captain Vikram Batra, PVC. http://www.bharat-rakshak.com. [Online] http://www.bharat-rakshak.com/LAND-FORCES/Army/Galleries/Courage/Batra/ (accessed July 23, 2014).

(Box continued)

> allowed his subedar to rescue the junior officer and step aside, but he took the initiative and his code of ethics made him think of his motto not only to safeguard the country but also consider the welfare of his co-soldier. Here Vikram maintained his ethical standards to live up to the army values which include personal courage and selfless service to the people and the country.
>
> Vikram Batra was a legend in himself and he proved that heroism is much more than just adulation and celebration. He acted ethically by taking responsibility and voluntarily going forward even in situations of extreme distress. His ideology and heroism sets an example for all the future leaders on how to lead by taking up responsibility when it matters the most.

3

BUILDING TEAMS: CREATING A PERSON–ORGANIZATION FIT

A team is a group of selected individuals, who are appointed on account of their expertise, skills, and attitude and are incorporated with the organization's values and mission. Each of the team members is assigned a given set of tasks and all of them work toward achieving their respective tasks, which lead them to achieve the common goal of the organization.

Building teams is a part of organizational development. It is looked upon as a source of effective functioning of the organizations. There are many dynamics attached to building teams in an organization. One has to understand how teams would function in a synchronized way to achieve the common organizational goal. In addition, one has to look at the team from all aspects, ranging from inter-group relations, formation of teams, groups, or pairs, and individual satisfaction as well.[1]

Thinking of team building in the military, one would think of an organization that has a demanding organizational structure with teams/units that are regimented. It follows a structure that has a

---

[1] Rothwell, W. J., & Sredl. H. J. (2000). *The ASTD reference guide to workplace learning and performance: Present and future roles and competencies* (3rd ed., Vol. 1). Amherst, MA: Human Resource Development Press.

top-down mechanism and not a collaborative one. The top-down mechanism indicates a decision-making process where the subordinates are not consulted or asked for inputs. However, the military still holds the coveted title of having the strongest teams across the globe. Yin-Che Chin, Yun-Chi Chen, and Ya-Lunstao proposed the following definition of organizational development:

> Organization development is a planned, system-wide, organization-wide, and values-based collaborative process of applying behavioral science knowledge and self-analytic methods to the adaptive development, improvement, reinforcement, technologies, markets, and challenge of such organizational features as the strategies, structures, processes, people, and cultures that lead to organization effectiveness and improvement.[2]

The aforementioned definition can well fit in today's requirement in an organization's development. Today, team building has gained a lot of attention in the research on development of organizations. Many new studies have indicated that team building is the most significant and powerful intervention in organizational development.[3] As companies are expanding and growing with globalization, more and more reliance of the organization's performance is dependent on its employees. Employees, as a part of the team, are basically the executors of the organization's processes and missions. It has become imperative that the employees of an organization integrate and collaborate with each other to become a workforce that can deal with the presence of the organization on a global platform.

Team building not only enhances the problem-solving skills, but also improves a team's productivity and efficiency. It gives a sense of belonging to the employees and makes them feel that they are a part of the organization.[4] Organizations' productivity is dependent

---

[2] Chen, Yin-Che; Chen, Yun-Chi; Tsao, Ya-Lun. (2009). Multiple Dimensions to the Application for the Effectiveness of Team Building in ROTC. *Education*, *129*(4), 742–754.

[3] Ibid.

[4] Sanders, R. L. (1992). What they forgot to tell you in the O.D. workshop: Your job. *ARMA Records Management Quarterly*, *26*(3), 46–52.

on the positives of team building, which further help organizations in increasing their efficiency and productivity.

TYPES OF TEAMS

Team building is a process directed toward accomplishment of tasks of an organization. Gibb Dyer suggested the following types of team building.[5] These types of teams can be built depending on the organizational structure, type and intensity of the tasks, and organization's goals and missions.

- **Manager-Led Team**
  A manager-led team is of a classic form, where a manager assigns a task to his team members. The manager/leader decides the tasks and the goals for the team. He/she then assigns a task to a selected team and then guides and supervises his subordinates, enabling them to accomplish the task. All the subordinates have to report to him on a regular basis.
- **Self-Managing Team**
  In a self-managing team, the team members are given the authority to progress with the goals and tasks set up by the manager, and also to choose the methods to accomplish them. Here, the manager/leader does not interfere much in the team processes and believes in his team members. He therefore, gives them the freedom to choose their own ways of accomplishing the tasks.
- **Self-Directing Team**
  In a self-directing team, the team members are authorized to decide the means through which they want to accomplish the assigned task. This form of team building is practiced in organizations or domains where the assigned tasks are constantly exposed to various vulnerabilities and externalities, and quick decisions need to be taken.

---

[5] Dyer, J. W. G. (2005). Team building: Past, present, and future. In W. J. Rothwell & R. Sullivan (eds), *Practicing organization development: A guide for consultants* (2nd ed.). San Francisco, CA: John Wiley & Sons, Inc.

- **Self-Governing Team**
  In this form of team building, the team members are given the authority to decide the tasks and goals of the team. These members also have the ability, power, and the authority to direct the whole organization. An example for this form of team building would be a team of board of directors. They monitor the performance of other teams and the organization as a whole. This type of team building can be chosen on the basis of the organizational process and the task assigned.

TEAM COMPETENCE INDICATORS

Dyer and Schein together developed indicators of team competence for a superior performance of a team.[6] They are listed and explained as follows:

- **Setting Clear and Measurable Goals**
  Setting a goal and communicating it to the team clearly is an essential part of team building. Setting up clear goals helps the team members in understanding the final aim and the expected result of the task assigned. It helps them in looking out for various means and directions to solve the arising problems and difficulties. Setting clear goals is like assigning a target to the team. Goals keep the team members motivated and ensure that they do not lose focus in the process. Goals give the team a sense of challenge and make them find new solutions to accomplish the task.
- **Building Clear Assignments**
  Allocating assignments is the duty of the manager/leader. He/she has to choose the right people for the task, and also give them a roadmap to follow. Assignments are basically subtasks of the main task allocated to the team. It is a set of steps and guidelines that helps a team in accomplishing the task assigned to them. Assignments clearly indicate the roles

---

[6] Dyer, W. G., Dyer, J. H., & Schein, E. H. (2007). *Team building: Proven strategies for improving team performance* (4th ed.). San Francisco, CA: John Wiley & Sons, Inc.

assigned to each of the team members and the stipulated time to accomplish the same. These assignments are well documented for further reference.
- **Implementing Effective Decision-Making Process**
  Depending on the nature of the task, the manager/leader has to identify and implement an effective decision-making process. Decision-making process should contribute toward building strong teams, and this can only be accomplished if decision-making process is collaborative in nature. In this process, all the members are allowed to raise issues, give opinions, and make suggestions. This process gives the leader a broader perspective of the issues that can be faced, and it also builds an environment where team members can come up with innovative ideas. It not only increases the efficiency of the team, but also gives them a sense of belonging, making them feel important and valued.
- **Establishing Accountability**
  The manager has to establish accountability of a team as well as that of its individual members. Building accountability is basically assigning responsibility to the team to achieve the task and to the individual members for their respective sub-tasks and roles. Establishing accountability at these micro and macro levels increases the performance standards and makes an efficient team. This is the effect of the synergy of the team and individual efforts to achieve the desired results. Building accountability also prevents individuals from blaming each other and makes them take charge of their own as well as group behavior.
- **Conducting Effective Meetings**
  Conducting regular meetings with the team members is a tool for effective team communication. Meetings give opportunity for open interactions and discussions among the team members and the supervisors. This gives supervisors an opportunity to understand their subordinates better and vice versa. Also, it brings more clarity to the process that has to be followed and leads to conclusions that can improve the problem-solving ability of the team members and the team.

- **Building Trust**
  Building trust in the team members is one of the strongest elements of team building. The team members should believe in each other, and also in their supervisors. If all the team members trust each other, it creates a feeling of reliability and unity among them. This can be further enhanced by explaining the employees the values, purpose, and mission of the organization. This should be followed by embracing everyone within the organizational behavioral standards.
- **Establishing Open Two-Way Communication**
  An open two-way communication indicates free communication between the leader and the team members. A lack of two-way communication can cause miscommunication and communication gaps among the team members. This will prevent the leader from understanding the problems faced by the team members and will make the leader less approachable, creating a gap between the minds of the members and their leader. Effective two-way communication can be attained through frequent team meetings and feedback forms.
- **Dealing with Conflicts**
  Every team experiences some friction and conflict among the team members. This is obvious when different minds from different backgrounds with different dynamics come together to achieve a common goal. However, sometimes, conflicts if not managed properly by the manager, can destroy the team's harmony and cohesion. The leader of the team has to make sure that he does not neglect any conflicts and tries sorting them on a priority basis. Anyone found breaching or sabotaging the team's unity should be given a strict warning followed by a strict action.
- **Ensuring Mutual Respect and Collaboration**
  Creating mutual respect and collaboration among the team members can be done by making them focus on the positive points of everyone. Members must know each other's strengths, creating a sense of respect for one's unique ability. This also motivates the members to help each other to accomplish their tasks. Mutual respect and collaboration among the team members build a strong team.

- **Encouraging Risk and Innovation**
  Organizations and leaders should encourage team members to undertake risks. Risks could be in the form of implementation of good ideas of the employees. This leads to developing an environment conducive to innovation and creativity. Innovation and creativity increase the efficiency and the performance of the team, as well as the individuals.
- **Engaging in Team Building**
  Team members can be engaged in team building by connecting them at the task level (professional level) and at the relationship level (personal level). To enable the members connect at a task level, open discussions and suggestions on achieving the common goal should be encouraged. Further, task assignments should be made on the basis of an individual's expertise, skills, and interests. To connect the team members at a personal level, leaders should encourage them to listen to each other and help them understand each other better.

SELECTING THE RIGHT EMPLOYEE

In today's market characterized by cut-throat competition, every company wants to keep the situation in control by attracting and retaining competent and qualified personnel. Behery identified four different criteria for choosing people who fit well with the organization. The following criteria must be looked for in an individual before appointing him/her:

- Congruence between individual and organizational values
- Goal congruence with organizational leaders or peers
- Match between individual need or preferences, and organizational systems and structures
- Match between individual personality characteristics and organizational climate[7]

---

[7] Behery, M. H. (2009). Person/organization job-fitting and affective commitment to the organization: Perspectives from the UAE. *Cross Cultural Management: An International Journal, 16*(2), 179–196.

Organizations can build an efficient team by getting the right people at the right place by considering the aforementioned points.[8]

- **Process of Hiring and Selecting Employees**
  In order to get success, persons recruiting the employees should seek potential candidates who can perform their jobs effectively and have the attitude of being committed to the organization. Hence the recruiters, while selecting, should try to understand person–organization fit and look beyond the traditional approach of matching the knowledge, skills, and abilities (KSAs) of an individual with the job attributes. Hence, laying down accurate selection and hiring criteria become crucial in order to take correct decisions. It will also enable the organization to predict the turnover risk before the individual is hired.

  Further, researchers such as Kristof-Brown, Zimmerman & Johnson mentioned the need to make the employees aware of the values and culture of the organization before designing the screening process. This clarity will help the organization in selecting those who would adjust and share these values and culture, and in screening out the rest.[9] *Organizations should try to use standardized metrics for assessing the employees.*[10] Organizations should ensure that the employee is comfortable in the organization and has the requisite abilities to be able to contribute to the organizational growth, and also accomplish his personal goals.[11] With the changing trends

[8] Sutarjo. (2011). Ten ways of managing person-organization fit (P-O fit) effectively: A literature study. *International Journal of Business and Social Science, 2*(21), 226–233.

[9] Kristof-Brown, A. L., Zimmerman, R. D., & Johnson, E. C. (2005). Consequences of individuals' fit at work: A meta-analysis of person-job, person-organization, person-group, and person-supervisor fit. *Journal of Personnel Psychology, 58*, 281–342.

[10] Carless, S. A. (2005). Person–job fit versus person–organization fit as predictors of organizational attraction and job acceptance intentions: A longitudinal study. *Journal of Occupational and Organizational Psychology, 78*, 411–429.

[11] O'Reilly, C. A., & Pfeffer, J. (2000). *Hidden value: How great companies achieve extraordinary results with ordinary people.* Boston, MA: Harvard Business School Press.

in the employees, different weightages should be assigned to different domains to identify the person who fits the best. Different evaluations will indicate different aspects of the employees' skill.[12]

- **Efficient Communication During the Selection Process**
 Organizations, while recruiting a candidate, should make every attempt to provide candidates with information regarding the training opportunities, level of responsibilities, values, culture, and policies to enable the candidate in assessing his/her fit in the organization. Organizations which are able to induce the perception of being a caring one tend to have a positive impact on the individuals. This idea was supported by Carless who believed that if an organization is able to display willingness in assisting an individual in choosing a job that will match his/her KSAs, values, needs, and attitude, it will be perceived as a caring organization in the mind of the applicant.[13]

- **Socialization**
 In spite of how well an organization has done during the recruitment and selection process, it is equally important for the organization to ensure that a new employee mingles well with everyone else and is able to adjust to its culture. New employees entering the organization are unfamiliar with its culture and thus, it is the duty of the organization to familiarize him/her with its culture. This process of adaptation is known as *socialization*. The process of socialization is also about transmission of values, attitudes, and assumptions from the senior employees to the new ones. Through the process of socialization, an organization ensures that the fit between the individual and the culture is smooth and comfortable. As per the views of Tepeci & Bartlett, the socializing process enhances the employees' belongingness for the organization,

---

[12] Sekiguchi, T. (2007). A contingency perspective of the importance of PJ fit and PO fit in employee selection. *Journal of Managerial Psychology, 22*(2), 118–131.

[13] Carless, S. A. (2005). Person–job fit versus person–organization fit as predictors of organizational attraction and job acceptance intentions: A longitudinal study. *Journal of Occupational and Organizational Psychology, 78*, 411–429.

which in turn is reflected in their satisfaction levels and the intention to remain in the organization for a longer time.[14]

- **Providing Comprehensive Training**

  As per Greenberg & Baron, training is the process of systematically teaching the employees to acquire and improve their job-related skills and knowledge to improve their job performance.[15]

  Well-structured training programs, which enhance the employees' KSAs to perform a task, help employees in adjusting and accommodating with the organization. Training programs should be designed in such a way that they complement the organization's strategic goals. This will help the employees in performing better and satisfying the needs of the organization.[16]

  Training can take two forms: formal and informal. In informal training, the older employees teach the new employees to perform tasks under their supervision. The other form of training involves formal training, where the employees are formally trained through deliberate and systematic efforts to impart them the skills required to accomplish the assigned task. However, the scope of training is not limited to new employees only. Even the older employees are made a part of the training to enhance and upgrade their skills and knowledge. Upgradation of skills is required to meet the new challenges that accompany the dynamic corporate environment.

- **Career Planning and Development**

  Career is the evolving sequence of various work experiences over a period of time. However, with time, individuals reach the peak of their careers and are likely to remain there, indicating stagnancy in their careers. At this point, the managers

[14] Tepeci, M., & Bartlett, A. L. B. (2002). The hospitality industry culture profile: A measure of individual values, organizational culture, and person-organization fit as predictors of job satisfaction and behavioral intentions. *International Journal of Hospitability Management, 21*(2), 151–170.

[15] Greenberg, J., & Baron, R. A. (2008). *Behaviour in organizations* (9th ed.). Upper Saddle river, New Jersey: Pearson Prentice-Hall.

[16] Autry, C. W., & Wheeler, A. R. (2005). Post-hire human resource management practices and person-organization fit: A study of blue-collar employees. *Journal of Managerial Issues, 17*(1), 58–75.

should intervene and help such employees in climbing the career graph and achieving their goals. For this purpose, the managers should invest in various human resource (HR) tools required for the development and career management processes for candidates with high potential, which in turn will help the organization in optimizing the employees that suit it the best.

- **The Role of the Leader**
A leader is a person who has the ability to influence others and seek their voluntary commitment toward the achievement of the group goals, preferably with enthusiasm. The interactive style of leadership has a significant positive effect on the employees' motivation, commitment, and trust in the leader, in spite of the fact that all of them are not the best fit for the organization.[17]

Further, it has been viewed that the leaders, through their communication and behavioral pattern, can influence the perception of their followers, which would ultimately influence the perception of the entire workforce.[18]

BUILDING THE TEAM

Building a team involves more than putting a group of people together. There are several indigenous and exogenous factors that need to be taken into consideration. It begins with choosing the right people based on their skills, merit, and suitability. Then, the team has to be nurtured with guidance, training, and supervision.

Every organization, irrespective of its function or location, has some targets and goals. The accomplishment of these tasks requires the combined efforts of a group. A group wherein all the members work toward accomplishing a common goal constitutes a

---

[17] Li, Ji. (2006). The interactions between person-organization fit and leadership styles in Asian firms, an empirical testing. *The International Journal of Human Resource Management, 17*(10), 1689–1706.

[18] Van Vianen, A. E. M., De Pater, I. E., & Van Dijk, F. (2007). Work value fit and turnover intention: Same-source or different-source fit. *Journal of Managerial Psychology, 22*(2), 188–202.

team. The team-building process and team structure vary depending on the organization and its demands. An organization would require a team with a leader and a board which handle the important matters. Army, on the other hand, is a large team, in which several smaller teams exist with officers of different ranks as leaders. A team can be regarded as the basic structural unit of any organization. Team building requires coming together of people from different backgrounds, the accomplishment of an organization's goal being the driving factor. Team building enhances the team productivity and efficiency; hence, selection of the most suitable candidate is essential for the army.

Team formation requires time. It takes time to take over from a former team and perform at their level right from the scratch. However, proper guidance, vision, and careful planning can help a new team in performing better.

- **Selecting the Right Team Member**
  Effective team-building activity starts with selecting the right members for a team. Members for a team are selected on the basis of the needs of the team, the organization, and the knowledge, skill, and attitude (KSA) of the individual. The selector, while appointing an individual, should also keep in mind the employee's compatibility (in terms of behavior and interests) with the other members of the organization, particularly the team members. An incompatible individual can disrupt the harmony and synchroneity of the team. On the contrary, a right member will integrate with the team easily. This helps the members connect at a professional and at a personal level, making the team cohesive.
- **Phases of Developing a Team**
  Once the leader has chosen the right people for a team, he/she has to work toward developing it. There are some basic stages involved in team formation which can increase the efficiency and productivity of any team, irrespective of the team or members. Five phases of development (Figure 3.1) have been laid down by Scott Tomek. These phases of development are described as forming, storming, norming, performing, and adjourning.[19] These terms were originally

coined by the psychologist Bruce Tuckman in 1965. The first phase indicates the stage where all the members of the team are assigned their respective roles, responsibilities, and team tasks and goals. The second phase is explained as the phase of conflict, where the members are trying to adjust to the new members and surroundings. This is then followed by conflict in the minds of the new joinees that they are not in the team. Soon, this phase is replaced by the third phase of norming. In this phase, the members of the team start communicating with each other, and begin solving the task-related problems and issues. This is the stage where the team actually begins to emerge. This phase is followed by the stage of performing.

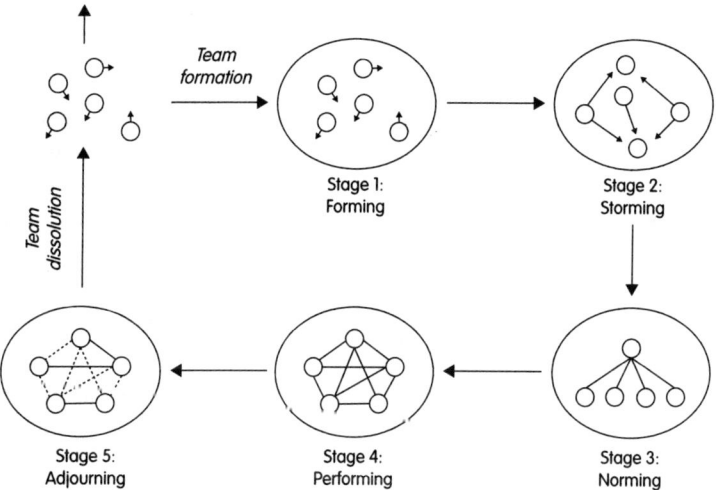

Figure 3.1 Stages of Team Formation

Source: Stages in Team Development (Adapted from Robbins, 1996 as cited in Wienclaw, 2014).[20]

[19] Tomke, S. (April 2011). Developing a multicultural, cross-generational, and multidisciplinary team: An introduction for civil engineers. *Leadership and Management in Engineering, 11* [Special issue], 191–196.

[20] Wienclaw, Ruth A. (2014). Teams & Team Building. *Research Starters Academic Topic Overviews*, January Issue, pp. 1–6, Retrieved from http://eds.b.ebscohost.com/eds/pdfviewer/pdfviewer?vid=5&sid=01f8e55b-ed08-4807-ac54-352996fb3a80%40sessionmgr113&hid=107 (accessed on July 25, 2014).

In this phase, the team members begin to work together and show a commitment toward the accomplishment of common organizational goals. This is the phase when a group of selected people turn into a team. In these four phases, those who do not fit in move out of the team and the rest who are on the plane stay. Phase five, that is, adjourning, indicates the final stage of accomplishing the task. This will involve making project reports, distribution of products, etc. In this phase, once the task is accomplished, the team disbands and the bond between the team members dissolves. The members then move back to the normal activities that do not involve the entire team.

As a team leader, your main aim should be to help your team reach and sustain high performance. The following steps will help ensure that you are doing the right thing at the right time:

- Identify the current position of your team, that is, at which stage of the team development your team is—from the categories mentioned above.
- Now consider what steps should be taken to move toward the performing stage and what you as a leader can do to help the team do that effectively. Table 3.1 helps you understand your role at each stage and reflect on how to move the team forward.
- Schedule a regular review of your team's position, and modify your behavior and leadership approach to suit the stage your team has reached.

- **Building Multicultural Environment**
  With globalization, organizational cultures are becoming more and more diverse and multicultural. This indicates that people from different walks of life, from different countries, and from different religions have come together, and are working in the same teams/groups. A multicultural environment is a work culture that fits in all the people from diverse backgrounds and uniqueness, with the same ease and comfort. No organization can avoid this today. In this scenario, there are chances of conflicts that can arise from individual differences.

Table 3.1
Leadership Activities at Different Group Formation Stages

Stage	Activity
Forming	Give directions to the team and establish the objectives clearly.
Storming	Establish correct process and structure, and work to resolve conflict, and build harmonious relationships among the team members. To improve performance of the team, and provide extra support to the weak and less secure members. It is important to explain the "forming, storming, norming, and performing" idea, so that people understand why conflicts occur and understand that things will get better in the future. And also, teach assertiveness and conflict resolution skills to the team, where necessary.
Norming	Step back and let the teams take responsibility of progressing toward the goal on their own. The best way is to arrange for a social or a team-building event.
Performing	Assign responsibilities to the team as sensibly as you can. Once the team has achieved high performance, the support provided to the team should reduce significantly. The leader should start focusing on other goals and areas of work.
Adjourning	When the task is accomplished, celebrate its achievement with the team members.

Source: Anonymous.[21]

These differences have to be considered by the authorities before the team has been made. These differences can be based on religion, language, culture, style, etc. Another basis of difference among the members can be the work culture, practices, ideologies, management structure, etc. All of these factors should be assessed by the leader before forming the team. Through his/her behavior, the leader should also demonstrate respect toward people coming from different backgrounds, and influence the team members also to do so.

- **Integrating Multiple Disciplines**

  Organizational teams today comprise of people from different disciplines. Organizations can have engineers, scientists,

---

[21] Anonymous. (n.d.). *Forming, storming, norming, and performing: Helping new teams perform effectively, quickly.* Retrieved from http://www.mindtools.com/pages/article/newLDR_86.htm (accessed on July 12, 2013).

and people from the humanities in one common team. All the members coming from different disciplines have different views toward different aspects of the task. The leader has to make sure that he/she helps the members overcome their inter-disciplinary differences. This can be done by defining the roles of each member clearly based on the member's expertise. Each member can be consulted for the aspect of the task he/she is in charge of.

- **Improving the Work Climate**
  Work climate of an organization is an important determining factor for building efficient teams. The armies today have got along with the current trends and have improved their organizational climate. They have implemented behavior-based performance indicators and appraisals. Training for everyone, be it a man or a woman, has been standardized to employ people on fair and just grounds on the basis of merit. Organizations should make provisions for giving feedback on the members' performance, telling them where they could improve on. Counseling should also be provided to guide people to make the right career decisions on the basis of their abilities and interests. Another important team-building factor is giving equal opportunity to everyone. Opportunity is given in the form of training, facilities, and examination. Giving equal opportunity to everyone ensures that the fittest and the most deserved people are taken up in the team. This breaks the barrier of any prejudice from the leader. This lures people to perform better and put in more efforts to improve their performance. This brings people with the same abilities together, creating efficient teams.

TRAITS THAT ARE CRUCIAL TO TEAMS

Lynda Gratton and Tamara J Erickson in their study, "Eight ways to build collaborative teams," suggested four traits that are crucial to a team.[22] These traits consider the changes that have occurred in

---

[22] Gratton, L., & Erickson, J. T. (November 2007). Eight ways to build collaborative teams. *Harvard Business Review*. Prod. #: R0711F-PDF-ENG, 1–11.

the team formation in organizations over years, indicating the new team-building scenario.

- **Large Team Size**
  Decades ago, sizes of teams in organizations rarely exceeded the number of 20 members. However, today this figure has averaged to approximately 100 members or more in a team. This significant increase has occurred because of the growth and expansion of organizations, which require more people to support it. There is the need of more people to maintain customers, coordinate with other stakeholders, etc. This study indicated that as the number of members increased from 20, the cooperation among the members decreased. Thus, this study indicates a negative relationship between the increase in the number of members and the cooperation among them.
- **Diversity**
  Organizations are required to select the best talent to fight the competition with other companies, along with expanding their market reach. To achieve this, the best talent is hired from different countries across the world. Thus, globalization has made teams diverse. People in one team can have different backgrounds and perspectives. This diversity has proved to be conducive to innovation in the organizations. However, research indicated a contrast: if people in a team have never met before and have come from different backgrounds, their likeliness to share knowledge with each other is less.
- **Virtual Participation**
  Most of the complex and collaborative teams today are functioning from a distance, via the web. This implies that they might not share the same geographical boundaries. This is because, for the business firms to operate across the globe, insights of members located in different places are required to coordinate and communicate. This basically dissolves the purpose and the importance of having a team use the same office. However, research indicated that as teams became more virtual their collaboration decreased.

- **High Education Levels**
  Companies employ people with the best education, meeting their standards. However, Gratton and Erickson's research indicated that if all the highly educated experts were placed in one team, the likeliness of the disintegration and unproductive team work increased. This indicates that hiring the best minds and putting them together does not necessarily lead to more productive, efficient, and collaborative teams.

IMPORTANT CHARACTERISTICS OF TEAM BUILDING

Beckard in his work, "What they forgot to tell you," suggested the following four recommendations in assessing the important characteristics of team building in the development of an organization:

- Establish and/or clarify goals and objectives
- Determine and/or clarify roles and responsibilities
- Establish and/or clarify policies and procedures
- Improve interpersonal relations[23]

FACTORS THAT DETERMINE A TEAM'S SUCCESS

There are many factors that determine a team's success. Maintaining relationships with the members of the organization, relationship with the supervisors/managers, training modules, organizational culture and environment, etc. play an important role in building a team's success rate.

- **Building and Maintaining Social Relationships**
  A study undertaken by Gratton and Erickson indicated that teams that performed complex tasks and were more innovative in an organization invested more in building and maintaining relationships across the organization. A peculiar aspect that the research pointed out was that, while building and maintaining

---

[23] Beckhard, R. (1969). *Organization development: Strategies and models.* Reading, MA: Addison-Wesley Publishing Company.

relationships with others, successful organizations used practices that were memorable and difficult for others to copy.[24]

- **Learning from Seniors**
  It is a proven fact that people can learn better from their seniors by observing and spending time with them. Very few employees from thousands actually get this opportunity to learn by observing the seniors in an organization on a regular basis. Gratton and Erickson's study also proved that seniors in the organizations had a significant impact on the new teams to be. The senior team member's performance has a trickle-down effect across the organization. If they get the opportunity to learn from the seniors, they learn much more, as the learning atmosphere is informal.

- **Mentoring and Coaching on Daily Basis: Developing a Gift Culture**
  To form a better team, executives must ensure that they obtain mentoring and coaching on a regular basis. Informal or less formal sessions and conversations with the mentors increase the collaboration among the team members, and other members too. This has been addressed as a gift culture by Gratton and Erickson. This is because the executives receive the most precious part of their senior's life, that is, their time.

- **Ensuring Requisite Skills**
  It has been observed that many teams followed a collaborative culture, but were not really skilled or capable of collaborating with the other team members. The team culture was collaborative because the organizational culture and the seniors encouraged them to do so. However, training executives to be skilled proved to increase the collaboration among team members. To be skilled, executives need to posses the qualities of engaging in productive and purposeful conversations, being able to appreciate others, and being able to resolve conflicts and issues creatively, etc. By training people to do so, organizations can increase team collaboration, and thus increase the team performance.

[24] Gratton, L., & Erickson, J. T. (November 01, 2007). Eight ways to build collaborative teams. *Harvard Business Review*. Prod. #: R0711F-PDF-ENG. 1–11.

- **Developing a Sense of Community**
  Organizations can develop a sense of community in their executives by organizing various activities that can bring people together. Cooking on weekends, coaching for a sport, or developing policies that encourage interaction at this level should absorb some investment. The HR teams in organizations are closely associated with these practices and help in creating a family-friendly environment. Organizing employee and family-friendly events also add to the fun element to work, and increase the opportunities where executives can meet and communicate with other members at a nonprofessional level. Organizations need to reinforce the idea that organizations or offices are not mere workplaces, but are more like a community.
- **Appointing the Right Team Leaders**
  Teams with high collaboration clearly indicate a positive influence of the team leader. Team leaders make a significant difference to the team collaboration levels. Team leaders have to be task oriented to ensure that the assigned tasks and the organizational goals are achieved. Similarly, it is also important that leaders are also relationship oriented. This means that leaders should not only be concerned about the performance of the team with respect to accomplishing the tasks, but also about the relation they share with the team and the team members share with each other. Gratton and Erickson observed in their study that majority of the teams that were efficient and successful were being led by a leader who was task as well as role oriented. Both these qualities should be assessed in the leader before appointing him/her.
- **Building Heritage Relationships**
  Heritage relationships indicate relationships between people who come from diverse and different backgrounds in a team. It was found that such team members find it difficult to interact and share knowledge with each other. It has been observed that team members who knew each other and were not strangers formed a better collaborated team, right from the beginning. It is important that the companies have well-established networks

across countries. They should also try introducing the team members before they join the team, so that when they meet they know something about each other, and can also connect over their common interests.

- **Role Clarity and Task Management**
  Team members should be assigned roles very clearly. They must be very clear about what they have to do and how to do it. This gives the team members the opportunity to come more prepared with their roles. Studies have proved that when roles are clearly defined to the team and the team members, they collaborate better. The clarity in the assigned role makes the members feel that they can perform significant amount of the task by themselves. Without clarity of tasks, team members tend to waste a lot of time trying to figure out things and making negotiations. It was also observed that if the path of achieving the team's goal was left vague, it brought more collaboration in the team.

Team building has become an important and an integrated part of organizational development today. Corporate organizations are trying hard to keep their employees happy and content in every aspect. Building teams is a determining factor for an organization's performance and success. The happier and satisfied the employees are, the better they will perform at work. Moreover, effective team-building practices lead to lowering of the employee turnover ratio and absenteeism. This increases employees' job retention, performance, and also builds a friendly relationship with the organization.

More and more organizations are making huge investments in building personal relationships with employees, creating an environment and culture, and creating a better workplace. The senior members play an important role in developing effective teams in an organization. Team building is becoming more and more important today, as organizations and their employees need to deal with various diversities, long-distance coordination and communication, build global teams that can deal with challenges encountered in the globally established and spread businesses.

> **TEAM WORK: BUDDY SYSTEM IN ARMY**
>
> *Every day I wake up, I have to thank Godding*
> *He put the rifle on his crown triggered the weapon on his head.*
> *He was surprised that he survived to see the next sunrise.*
> *He was surprised that he survived the desired encounter with death.*
>
> Soldiers are always hard like a stone, fast like a wind but when it comes to their loved ones they are as soft as a feather.
> It was a hot Iraqi summer of August 2008 when Spc. Joe Sanders was deployed to Iraq. He was already stressed after breaking up with his wife and yet he had several months of deployment ahead.[25] In August 2008, when Spc. Joe Sanders felt that the journey of his life was coming to an end, he wanted to commit suicide. Spc. Joe Sanders was extremely stressed due to a strong feeling of isolation from his family. Corporal Godding was Sander's buddy. Godding noticed signs of stress in Sander's action and removed the firing pins from his weapon the day before he attempted suicide. On that fateful day when Sander's pointed his gun to his temple and pressed the trigger, the gun did not fire. Godding was awarded Meritorious Service Medal by the US army for exhibiting team spirit.
> The aforementioned incident typifies the importance of buddy system in the army. Buddy System is an arrangement in which two teammates are paired based on their compatibility, assistance and teamwork as well as friendship. Buddy system is an integral part of many armies of the world and seeking help from your buddy is not considered to be a weakness, but is encouraged. Soldiers are often away from their friends and families when deployed in non-family stations or/and conflict zones. Isolation from one's family and friends could result in extreme stress. Sander once quoted "Fighting doesn't bother
>
> *(Box continued)*

---

[25] Morgan, Z. Soldier saves battle buddy's life with simple act of caring. http://www.army.mil. Retrieved from http://www.army.mil/article/39171/ (accessed on May 14, 2010).

(Box continued)

soldiers, we do that all day long. What bring stress in field is being away from our loved ones."

Most army operations are small team operations. Small team needs synergy and cooperation to result in good performance. Buddy system affords creation of synergy and cooperation, which not only eases the level of stress among teammates but also creates camaraderie in the small teams. Buddy system demonstrates the eminence of teamwork that fuels common people to achieve uncommon results. Through buddy system, teammates share their thoughts and feelings with each other allowing them to develop a relationship that facilitates cooperation, reciprocity, respect, and mutual understanding. When the going gets tough, extra motivation from one's buddy goes a long way in ensuring organizational success and personal well-being.[26]

---

[26] A victory for Army's war on suicide: Soldier saves a buddy from taking his own life. http://www.foxnews.com. [Online] http://www.foxnews.com/us/2010/06/19/victory-armys-war-suicide-soldier-saves-buddy-taking-life/ (accessed on June 19, 2010).

4

WORKFORCE MOTIVATION

The U.S. Army Handbook (1973) explained motivation in an army as "[a] person's motivation is a combination of desire and energy directed at achieving a goal. It is the cause of action. Influencing someone's motivation means getting them to want to do what you know must be done."[1] Motivation can be defined as the thrust that propels a person to some action or behavior. The word "motivate" means to provide reasons for action. Motivation, thus, can be explained as a reason for which one exerts to put effort. This motivation is derived from individual needs, wants, and drives.[2]

Motivation can be described as an individual's willingness and persistence to accomplish a particular task. It is generally dependent on two factors; one is the intensity of the need, and the second is taking an action so that the need can be satisfied.[3] The intensity of need can be verified with an example, wherein an individual needs to choose between two important needs. If the needs of these

---

[1] Handbook, U. A. (1973). *Military leadership*. US Government Printing Office.
[2] Timm, P. R., & Peterson, B. D. (2000). *People at work: Human behavior in organizations* (5th ed.). Cincinnati, OH: South-Western College Publishing.
[3] Anonymous. (2002). Motivational Principles and Techniques, U.S. Marine Corps. JROTC, Category 1 – Leadership, Skill 6 – Esprit De Corps.

two things do not overlap then the person is motivated to work. However, if the work is interrupting with his/her other needs then this will affect his/her motivation to accomplish the task and to give in more effort. The second factor can be explained as the opportunity cost. An employee will choose to prioritize between his vital needs, depending upon the opportunity cost he/she has to bear in letting go or compromising on one of them. For example, if one has to choose between work and home, one might have to choose work over home because of professional responsibilities. However, this reduces the individual's motivation to work.

The aforementioned situations are very common in today's world, as people have to really struggle to maintain the balance between work and home. Work–life conflict has been a major reason why employees' motivation at work has been found reducing. Dual career, working overtime, lack of personal, and uninterrupted time and space have led to an increase in work–life conflicts, which have a direct impact on the employees' motivation to work. Thus, motivation as a topic of research has gained a lot of interest lately. Many companies have experienced a decrease in their aggregate performance because of low productivity of the employees. However, companies have recently realized the importance of workforce motivation and are making efforts to keep their employees motivated.

Military, as an organization, is the best example for building high workforce motivation. This can be observed from the motivation soldiers have, for undergoing rigorous training, keeping away from family, and being exposed to extreme and uncertain situations, which can also cost them their lives. However, they are determined to stay and dedicate their lives in the military, as they are driven by constant motivation. This motivation comes in the form of job satisfaction that is derived from being a part of and contributing to public service. Members of the army are driven to perform their best because of the sense that they are safeguarding the national security. However, as one undergoes the military training, and faces disappointment and extreme stress, the public service motivation does fall down. This can prompt him/her to quit and move back to the civil life. However, to combat this, the military has a motivation system built in the organizational processes. The organization's trainings and other sessions are designed in a way that the officers are constantly reminded of their goals, and are pushed and guided to move forward. Factors such as performance appraisal, on the basis of meritocracy and equality, develop a sense of respect for the organization. This motivates the members to give in more efforts and contribute to the organization.

PROPERTIES OF MOTIVATION

On the basis of the various definitions given by experts, the following definition can be provided. "Motivation becomes those psychological processes that cause the arousal, direction, and persistence of voluntary actions that are goal directed." Based on this definition, Terence R. Mitchell tried explaining the properties of motivation:[4]

- **Motivation as an Individual Phenomenon**
  This means that motivation is individualistically customized. Motivation encourages people to come out with their uniqueness and use it for achieving the desired goals. This uniqueness can be in the form of different needs, wants, attitudes, goals, etc.
- **Motivation Is Intentional**
  This means that an individual is motivated about doing something when he really wants to do, and not because someone else wants him do it. This indicates that motivating oneself to the majority of the extent depends on oneself and it is in one's own control.
- **Motivation Is Multifaceted**
  Motivation has many aspects. Motivation is driven by elements such as energy, direction, choice, etc. Therefore, motivation is not a self-igniting force, but it is a force driven by many other forces.

---

[4] Mitchell, R. T. (1982). Motivation: New directions for theory, research, and practice. *Academy of Management Review*, 7(1), 80–88.

- **Motivation Is to Predict Behavior**
  Motivation is concerned about an individual's action, and the internal and external forces that influence it.

The aforementioned properties of motivation change the structure of the previous definition. Mitchell came up with the following modified definition: "Motivation becomes the degree to which an individual wants and chooses to engage in certain specified behaviors. Different theories propose different reasons, but almost all of them emphasize an individual, intentional choice of behavior analysis."[5]

MOTIVATION SYSTEM

Every organization should have a motivation system installed within it. However, there are certain factors that should be taken care of before doing so. Mitchell came up with the factors determining the motivation system. The first factor is a good performance appraisal system. A good performance appraisal system will ensure that the factors that cause changes in an employee's performance are known and measured. The second factor is that motivation should be considered as an important factor contributing to the increase in employee's performance. The third and the last factor is that motivation should be the largest contributor to employees' performance.

FACTORS CONTRIBUTING TO HIGH WORK PERFORMANCE

Motivation is considered as a partial contributor to employees' work performance. It is said to be directly related to an individual's performance level. Research indicates that if the motivation is high, employees' work performance is also high (Figure 4.1).

Mitchell came up with the following factors that contribute to the high performance of employees:

- Knowing what is required (role expectations)
- Having the ability to do what is required

[5] Ibid.

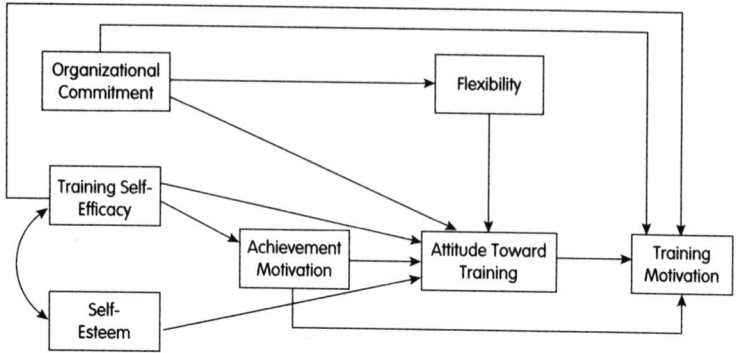

Figure 4.1 Antecedents of Pre-Training Motivation
Source: Author's own.

- Being motivated to do what is required
- Working in an environment in which intended actions can be translated into behavior.[6]

ORGANIZATIONAL CHALLENGES IN WORKFORCE MOTIVATION

In today's dynamic corporate world, recognizing factors that motivate the workforce has become a challenging task for the managers. This is because drastic changes have occurred in the workforce over the past few decades. Today, everyone is motivated by different factors. Previously, the employment scenario was such that the employees joined and retired from the same organization. They did not really consider changing their jobs to gain better job profiles and better remuneration. However, the employment dynamics have taken a drastic turn. Employees, today, are willing to explore different organizations and different domains to get better and more lucrative employment options. Employees are willing to pursue many careers in a lifetime. Hence, for a company to retain its competent employees, it has become imperative to keep them motivated and to understand the motivational factors of workforce.[7]

[6] Ibid.
[7] Javitch Associates. *Motivating employees.* Retrieved from http://www.javitch.com/Q/004.pdf

MOTIVATION IN ORGANIZATIONS: STATUS QUO

Motivation at today's workplace is affected by several factors that did not exist before. Companies have to continuously find new ways to keep their workforce motivated.[8]

- **Decreased Emphasis on Money**
  There was a time when leaders used monetary remuneration as a reward to motivate employees. Also, a common perception of a good job was a job that pays well. Money is definitely an important motivating factor at work, but only a good pay is not motivating enough for employees today. Money has proven to be a motivation booster in the short run, but is not motivating enough in the long run. To keep employees motivated in the long run, employees look for job satisfaction, work–life balance, work culture, etc. Organizations will have to work on these factors to keep the employees motivated.

  A Quarterly Survey conducted by McKinsey in 2009 concluded that the three non-monetary motivators—*getting praise from immediate managers, getting attention from leaders, and getting opportunities to lead task forces or to lead*—are no less effective, and, in some cases, are even more effective than the three highly rated monetary benefits—*cash bonus, increase in basic pay, and stock or stock options*. The results of the survey further throws light on the fact that the three aforementioned non-financial motivators play an important role in making the employees feel that they are valued by the company. This also ensures that the organization will take care of their well-being and give them copious opportunities for self-growth.[9]

---

[8] Lydia, B. (1997). *Motivation in the workplace: Inspiring your employees.* Virginia Beach, VA; Coastal Training Technologies Corp. Retrieved from http://www.trainingsolutions.com/pdf/motivating.pdf (accessed on February 27, 2014).

[9] Dewhurst, M., Guthridge, M., & Mohr, E. (June 2009). Motivating people: Getting beyond money. *McKinsey Quarterly.* Retrieved from http://www.mckinseyquarterly.com/Motivating_people_Getting_beyond_money_2460 (accessed on February 27, 2014).

- **More Amount of Work**
  With the cost-cutting measures adopted by the companies and technological advancement, the workload of the employees has increased manifold. This requires the employees to learn and adapt to new things, and, that too, promptly. Although these changes do offer many benefits, they come with the stress associated in learning them. This stress does affect employees' motivation to work.
- **Working in Teams**
  Organization is a group of people working together toward accomplishing the common objectives of the firm. No individual can work alone to accomplish a task. Employees have to come together and work in teams for the successful functioning of the organization. If individuals are able to cope with the team and build a strong bond as a team, they would be more committed to the organization. This, in turn, will prove to be a motivation booster for the employees. However, if the individuals are not able to get along with each other, small differences and issues can take the form of team conflicts or negative feelings among individuals about their performance. This would negatively affect the motivation of the team.
- **External Factors**
  There are certain factors, other than organizational and personal factors, that can contribute to individuals' motivation to work. Factors such as changes in the tax structure, which impose that more amount of their salary will be contributed to the government without an increase in the salary, can be a demotivating factor to work more. Economic instability, such as high recession rates that lead to high unemployment rates, can cause high stress, reducing their motivation to work. However, in rough economic conditions, employees are motivated to perform better because they are afraid of being laid off from the organization.[10]

[10] Perry, L. J., & Porter, W. L. (January 1982). Factors affecting the context for motivation in public organizations. *The Academy of Management Review*, 7(1), 89–98. Retrieved from http://www.jstor.org/stable/257252?origin=JSTOR-pdf (accessed on February 27, 2014).

MILITARY LEADERSHIP IN MOTIVATION: LESSONS FOR CORPORATE ORGANIZATIONS

The military operates in a top-down mechanism or in a leader-centric way. This means that all those employed in the military, irrespective of the level at which they are working, are guided and monitored by the leaders since their first day of training. This leader-centric approach in the military helps in developing effective leaders, who would be able to motivate their subordinates to perform efficiently, even in the toughest situations and conditions. The effectiveness of leaders in motivating the subordinates in the military is not effective in the corporate organizations. Reasons for this are mentioned as follows:[11]

- **Lack of Formal Training**
  The military follows a unique system of leadership. There are formal training programs for all the selected members in all the domains of the military. This system ensures that required training is given to the personnel, irrespective of their hierarchical positions. The qualities of leadership are drilled in the officers, even before one is assigned to perform the task and is followed throughout his career in the military. This enhances their leadership skills, which enable them to motivate themselves, and others too, when needed.
  However, this type of continuous drilling is not available in the corporate. Too little attention is given to the leadership skills possessed or to the potential to develop those skills, while filling the position of first-level supervisors. Also, no formal leadership training and feedback are given to the first-level supervisors afterward. This ultimately leads to sub-optimal supervision, which does not evoke high level of motivation.
- **Unit Ceremonies**
  Military units are engaged more in unit-level activities. This serves as a base in developing the team spirit of the unit.

---

[11] Wink, R. (2012). *How the military gets the motivation right.* FederalNewsRadio.com. Retrieved from http://www.federalnewsradio.com/?nid=395&sid=2749626 (on February 27, 2014).

One example of this is the ceremony of formal change of command. These ceremonies act as a great motivational tool for everyone in the unit, since these ceremonies enable the members of various units to learn from the experiences of the outgoing commander and the new aspirations of the new commander, his leadership philosophy and aspirations from the unit. Organizing unit ceremonies keeps everyone in the unit motivated to work for a common goal and purpose. Other unit activities such as physical training, and playing sports, etc. serve as an excellent opportunity to build unit cohesiveness and build motivation in the taskforce.

The aforementioned practice is not followed by most of the corporate organizations. Normally, when a manager or a supervisor leaves the organization and a new leader is appointed, employees are filled with feelings of apprehension, and not of positiveness. The phrase that best fits here is: "[A] known evil is better than an unknown one." Therefore, instead of feeling motivated, the employees feel a sense of insecurity and uncertainty. To adjust and accept a new leader, senior leaders are required to take charge of the situation, spending extra effort in stabilizing the environment, and keeping the employees motivated in the transition phase. Hence, it becomes imperative for the management of a corporate to arrange informal group events, such as get-togethers, concerts, competitions, etc., to keep them motivated.

- **Day-to-Day Emphasis on Leadership**
Leadership is an ongoing process. It requires experience, knowledge, decision-making abilities, cognitive abilities, etc. to create a leader. The military understands this progression. In the military, people are trained and taught to be good leaders on a daily basis. This throws light on how critical leadership is in motivating the taskforce. Corporate organizations also need to adopt the system of inculcating leadership in the team members. Leaders are not born that way. An individual has to be trained and taught to become one. Training and teaching should be done on a daily basis, as is done in the military.

Factors That Affect Training Motivation

Organizations should employ people who are willing to work, learn, and upgrade. A person's willingness to learn is an important factor that will determine his productivity and career growth prospects. Carlson, Bozeman, Kacmar, Wright, and McMahan have laid down a few factors that can help in assessing the individual's willingness to work and his/her contribution to the organizational development.[12]

- **Self-Efficacy**
  Self-efficacy is an individual's perception of his power and ability to achieve results.[13] People with high self-efficacy experience higher-level motivation for training than the ones with low self-efficacy levels. Self-efficacy is, thus, positively related to training and motivation.
- **Self-Esteem**
  Self-esteem is closely related to an individual's self-efficacy. Self-esteem is the likeableness for oneself in an individual's mind. The higher the self-esteem is, the higher will be the motivation for training, compared to those who have low self-esteem.[14]
- **Organizational Commitment**
  Organizational commitment has been explained as the intensity of an individual's identification and involvement in a particular organization.[15] Individuals who display higher degree of commitment indicate higher motivation for training. Thus, individuals with higher commitment should be sent for training.

---

[12] Carlson, S. D., Bozeman, P. D., Kacmar, K. M., Wright, M. P., & McMahan. (Fall 2000). Training motivation in organizations: An analysis of individual-level antecedent. *Journal of Managerial Issues, XII*(3), 271–287.

[13] Bandura, A. (1982). Self-efficacy mechanism in human agency. *American Psychologist, 37*, 122–147.

[14] Brockner, J. (1988). *Self-esteem at work: Research, theory, and practice.* Lexington, MA: Lexington Books.

[15] Porter, L., Steers, R. M., Mowday, R. T., & Boulian, P. V. (1974). Organizational commitment, job satisfaction, and turnover among psychiatric technicians. *Journal of Applied Psychology, 59*, 603–609.

- **Flexibility**
  Flexibility is an important factor for determining an individual's motivation for training. If an individual is more flexible, he/she will find it easier to adapt to new things and new environments. Thus, an individual's flexibility is directly related to his/her motivation for obtaining training.
- **Achievement Motivation**
  Achievement motivation can be explained as one's desire in achieving challenges and personal goals. Individuals who are motivated to achieve things in life have a positive attitude toward obtaining training, as they take up training as a source of overcoming shortcomings and learning new things, and, thereby, adding to their skill sets.
- **Attitude toward Training**
  An individual's attitude toward training can determine his/her motivation for training. His/her attitude can display the motivation for acquiring new skills and knowledge. Thus, an individual with a positive approach toward learning new things will display a positive attitude toward training.

Motivating Factors for Soldiers in Military

- **The Leader as Protector**
  A leader is the crucial link between the officers at higher positions and his subordinates in the military. A field commander should build such a rapport with his soldiers that they are able to identify with their commander, with the organizational values that he/she preaches, and with the missions that he/she is ordering them to accomplish. The unit leader should quite often display affection toward his/her subordinates by taking care of their physical and emotional needs. When leaders demonstrate adequate care to their subordinates, the soldiers feel secured, and, hence, carry out their duties diligently without the need for much supervision. Taking care of soldiers does not only mean comforting and protecting them, but also training them to become skilful soldiers who

could survive on a battlefield because they are technically, physically, and mentally proficient.[16]
- **Unit Cohesion**
Various research studies indicate that unity or *esprit de corps* not only strengthens a unit's level of morale, but also acts as "a powerful preventive measure against psychiatric breakdown in battle, and as 'generator' of heroic behavior among the unit's members." Additionally, "the normative power of the cohesive group causes the strong personal commitment on the part of the soldier that he ought to conform to group expectations."[17]
- **Mission Accomplishment**
The sense of achievement is one of the other factors that provide soldiers with high levels of morale and combat motivation. These are "for each soldier, a goal, a role, and a reason for self-confidence." In order to sustain a high level of motivation, a soldier must have a definite and tangible objective.[18]
- **Self-Confidence**
As a soldier proceeds with his training and experience in battle field, his role and self-confidence strengthen at the military operations. Training is a key ingredient for increasing or maintaining the soldier's combat morale, both at the individual and unit levels. Indeed, as S. L. A. Marshall noted, the "tactical unity of men working in combat will be in the ratio of their knowledge and sympathetic understanding of one another." However, no matter how much energy is put into training the soldier, if he/she is not adequately motivated, the outcome will constantly be low combat performance, because ultimately "performance equals knowledge times motivation."[19]

---

[16] Catignani, S. (2004). Motivating soldiers: The example of the Israeli defense forces. *Parameters, 34*(3), 108–121.
[17] Ibid., 116.
[18] Ibid., 117.
[19] Ibid., 117–119.

The Four Drives That Underlie Motivation in an Individual

Research studies propound that people are guided by four basic emotional needs or drives that are the products of our common evolutionary heritage. As set out by Paul R. Lawrence and Nitin Nohria in their 2002 book *Driven: How Human Nature Shapes Our Choices*, "they are the drives to acquire (obtain scarce goods, including intangibles such as social status); bond (form connections with individuals and groups); comprehend (satisfy our curiosity and master the world around us); and defend (protect against external threats and promote justice)."[20]

1. **The Drive to Acquire**
   We are all compelled to achieve goods that support our sense of well-being. We feel joyous and contented when this drive is fulfilled, and discontented when it is thwarted. This phenomenon applies to both physical goods like food, clothing, housing, and money, and to experiences such as travel and entertainment.
   "The drive to acquire tends to be relative (we always compare what we have with what others possess) and insatiable (we always want more)"; this explains why people always care not just about their own compensation packages, but about others' as well. It also illuminates why salary caps are hard to impose.[21]
2. **The Drive to Bond**
   Not only do humans bond with parents, kinship groups, or tribes, but also with larger collectives such as organizations, associations, and nations. When the drive to bond gets accomplished, it brings along strong positive emotions such as love and care, and when not, negative emotions such as loneliness and solitude develop. At work, the satisfaction of this drive boosts motivation, as employees feel proud of

---

[20] Nohria, N., Groysberg, B., & Lee, L. (2008). Employee motivation: A powerful new model. *Harvard Business Review, 86*(7/8), 1–9.
[21] Ibid., 2.

belonging to the organization and lose morale when the institution betrays them.[22]

3. **The Drive to Comprehend**
The drive to comprehend develops the desire to make a meaningful contribution. Employees working in various organizations are motivated by jobs that challenge them or initiate creativity, as they enable them to grow and learn, whereas they feel demoralized if work appears to be monotonous or leads to a dead end. Talented employees who feel trapped and bored often leave their organizations to find new challenges somewhere else.[23]

4. **The Drive to Defend**
It is a natural tendency of human beings to defend themselves, their property, accomplishments, their family and friends, and their ideas and beliefs against external threats. Fulfillment of the drive to defend generates a sense of security and confidence, and if not fulfilled, it produces strong negative emotions such as fear and resentment. The drive to defend tells us a lot about people's resistance to change.[24]

These four drives are independent of each other, and they follow no hierarchy or substitution for one another. Monetary incentives are not enough to motivate employees and to make them feel enthusiastic about their work in an organization. To fully motivate your employees, you must address all the four drives.[25]

HOW CAN FIRMS MOTIVATE THEIR EMPLOYEES?

A challenge that all the organizations are facing today is reducing the turnover ratio and increasing the employee retention rate. This can be achieved by keeping the taskforce motivated to perform well each time. Organizations will have to bring changes in their

[22] Ibid.
[23] Ibid.
[24] Ibid.
[25] Ibid.

framework, work environment, and work culture to match to the needs and wants of the employees today.

Redesigning Jobs

One way of keeping the employees motivated is by redesigning their jobs. Redesigning of jobs makes the work challenging and prevents it from being monotonous. Thus, when employees are enjoying their work, they will be willing to perform well and give more efforts to it.

- **Cross-Training**
  Cross-training is also known as job rotation.[26] Cross-training is the practice of shifting employees from one task to another. This practice can be adopted by corporate organizations when it is realized that the employees are suffering from routinization. Cross-training of employees is done normally on the same level. Employees get tired of doing the same thing again and again, as the job does not remain challenging anymore. Therefore, exposing employees to different and new challenges and activities prevent them from boredom, and keep them motivated. Job rotation also has positive results for the organization, as a wider range of skills in the employees gives the organization a wider scope for flexibility in terms of adapting to changes and being able to provide flexible work schedules.

- **Increase in the Variety of Number of Tasks: Job Enlargement**
  Judge and Robins have explained job enlargement as increasing the diversity of the job of the employees.[27] This means that as employees become more efficient with one task, more achievable tasks should be added to their work basket. The intention is not to increase the work load or burden, but to increase the scope of the job. This ensures

[26] Robbins, P. S., & Judge, A. T. (2011). *Organizational behavior* (12th ed.). Manipal, Karnataka: Prentice-Hall India.
[27] Ibid.

the employees that more valuable tasks are given to them as they are improving their performance, which keeps them motivated to perform better.
- **Vertical Expansion of Jobs: Job Enrichment**
Judge and Robins explained vertical expansion of jobs as a way of enriching the employees' work profiles.[28] Vertical expansion refers to the control and contribution one can have in the planning, execution, and evaluation. A vertically expanded job gives employees the liberty to complete an activity with full freedom, independence, and responsibility. This process is followed by giving feedbacks, which help the employees in assessing their own mistakes and drawbacks, and in improving on them.

Creating Alternative Work Arrangements

Providing employees with alternative work arrangements enables them to work when they are at their best, and also gives them the comfort and space to manage between work and home. When an individual is compelled to work during a time period when he/she is not willing to, because of other priorities, the individual is demotivated to work. Flexible work environment increases employees' motivation as they work when they want to work, and not out of compulsion.

- **Flexible Work Timings**
Flexible work timings have proven to be an effective tool for motivating employees. Everyone today is multitasking, trying to maintain balance, and fulfill their needs. To accomplish all this, one requires time. Flexible work timings can give time to a mother to take her child to school and then come to work. For someone who wants to study further, gets time to prepare for it, and someone who is passionate about hobbies will find time to do what one loves the most. All of these add to employees' satisfaction, and thus, keep them motivated to

---

[28] Ibid.

work better. This also gives employees a feeling that their organization understands their needs, and, thus, increases their commitment, reduces absenteeism, and, most certainly, increases motivation.

- **Job Sharing**
  Job sharing is a practice offered by organizations where a regular job can be shared by two people. They could either work part time each, or on alternate days, or on weekly basis. Research says that 31 percent of the organizations are offering this facility today. This not only provides the employees with work flexibility, but also gives the organization the opportunity to have two talents employed for a task. This can work best for woman employees. Also, if one of the employees is not present at work, the other can always make up for it, and also get paid. This works in favor of the employee as well as the employer. However, this practice has not gained much attention yet from the employees' end. One of the major reasons for the lack of acceptance for job sharing is because of not being able to find right partners for sharing the job, and the negative perception associated with a half badly done job, that is, if one of the partners does not perform a job well, the blame will have to be shared by both.

- **Telecommuting**
  Telecommuting for work basically means working from home, without making any travel to work, with flexible work hours and no need to dress up. While telecommuting, individuals are supposed to work for two days a week on the computer, which is connected to the network at work. Many companies and employees have opted for telecommuting. However, organizations can offer this on the basis of the nature of a job. Three categories that best fit this practice are: *1. routine information-handling tasks, 2. mobile activities, and 3. knowledge-related tasks.* This practice benefits the organizations as well the employees. It benefits the organization with an increased workforce, higher productivity, low employee turnover, reduced office costs, and increased motivation to work.

Increasing Employee Involvement

Increasing employees' involvement refers to an increased participation of the employees in organizational activities. This gives the employees a sense of commitment and responsibility toward work, keeping them motivated. This can be done by involving employees in the decision-making process, increasing the control on their work and lives in general. Organizations have employee programs built for this purpose; however, they are different in different countries, and also are dependent on the work domains.

One way of doing this would be *participative management*. Participative management is a program wherein the employees are given the power and the right to be a part of the decision-making process of the organization, via the supervisors. This has proven to be a good tool for boosting morale and increasing productivity. However, for this program to be successful, employees have to be equipped with the required knowledge and interest. Besides, there should be sufficient confidence between both the parties.

Another way of promoting employee involvement in an organization could be through *representative participation*. Representative participation is the most widely used employee involvement program in the world. In this program, instead of the employees participating in the decision-making process of an organization, a selected small group of two is chosen and they participate in the discussions and decisions.

Another program called *quality circles* is used to increase employee involvement in the organization. In this program, managers or supervisors from different domains form a team and are supposed to meet on regular intervals, say once a week, and discuss the problems and issues faced by different departments and find solutions. This is a good way of increasing employee involvement, as people from different domains can also contribute to other domains.

Rewarding Employees through Pay Programs

Rewarding employees is about rewarding them for their exceptional contribution to the organization. Reward programs prove to be motivating, as when people think about why should they be giving in more to the organization, the answer should be that the organization will pay them back for it, in cash or in kind. This can be done via variable pay and skill-based pay programs.

Variable Pay Programs

- **Price-Rate Pay**
  Price-rate pay is a pay program wherein a fixed amount of payment is made to the employees for a particular job. Depending on the work profile, people are either paid on the basis of the number of hours worked, or the number of targets achieved, or the number of goods produced. Therefore, the more an individual works, the more he/she gets paid in monetary terms.
- **Merit-Based Pay**
  Merit-based pay is on the basis of an individual's performance at work which cannot be quantified. If an individual has performed better than others, he/she gets paid on the basis of the increase in performance. This is an efficient program which differs from the price-rate pay, as those working more to get a higher pay will be eliminated. This gives employees the motivation to perform better at work.
- **Bonuses**
  Bonuses are huge amounts of money given to the employees at the end of the year, or during a festival, or during the end of the financial year, depending on the profits made. This program is very popular among the low-rank employees. It is a strong motivational tool that keeps them motivated to give in more efforts and stay committed to the company.
- **Share in Profit**
  Profit sharing is a pay program initiated by the organization to give all the employees or a selected few employees a share in the profits of the firm. Normally, high-level executives and managers are provided with this benefit. This is a strong motivating factor, as it indicates that, as the profits of the organization increase, the profit share also increases.
- **Gain Sharing**
  Gain sharing has gained a lot of attention lately. This is a variable-pay incentive program, where improvements in the

team performance from one period to the other determine the total amount of incentive the group will receive. Gain sharing increases team work, team spirit, team cohesion, and team motivation to perform better each time. It also brings a sense of healthy competition.

Skill Based Pay Programs

Skill-based programs aim at rewarding the employees for the skills they have. One is not identified by how much work one does, but is identified by what skill one possesses. It is an individual's skill that determines how well he/she is going to perform the task. It will also determine how he/she will contribute to the team and help others cope up. An individual's skill determines the longevity of his/her career and the progress in his career graph. Skill-based pay programs can be rewarded in the form of flexible benefits. Flexible benefits are customized depending on each individual and his/her needs.

Let us talk about the three most popular benefit plans: modular plans, core plus option, and flexible spending plans. Let us look at them one by one.

- **Modular Plans**
  Predesigned benefit programs are developed for a particular group/set of employees, matching their needs. The differentiation can be made on the basis of the marital status of an individual, number of children, number of dependents, etc.
- **Core Plus Option**
  In this benefit program, employees are given access to the essential benefits and they have to choose from a list of options. Each employee gets benefit credits, which can enable him/her to make purchases later. As an employee makes a choice from the list, he/she chooses the benefit that suits his/her needs.
- **Flexible Spending Plans**
  In flexible spending plans, organizations choose to pay for particular services, for example, paying for employees' health

insurance. This frees the employee from paying for his health, and, thus, increases his real salary.
- **Employee Recognition Program**
Employees who perform exceptionally well or better than the other team members should be given recognition for the same. Recognition can be in the form of an applaud by the team, a mere "thank you" by the manager, or even a "well done" by the manager. This form of recognition has a high motivating value, as it gives employees a sense of importance, a feel-good value, and makes work enjoyable.

A motivated workforce is the blood of an organization. It is important that the organizations make required investments and take appropriate initiatives to motivate their employees. A motivated team, in the long run, gives greater benefits to the organization. It not only increases the productivity of the organization at the micro and macro levels, but also saves a lot of cost of the company. A motivated team has higher commitment to the organization. This reduces the employee turnover, and, thus, reduces the costs of training and employing new staff members. Also, bringing the new employee to the efficiency level of the trained employee incurs a huge cost, in terms of the time required to do so.

Organizations also need to understand that money is not equal to motivation. Motivation is about giving the person what he/she deserves and helping him/her perform better. Motivation, besides providing monetary remunerations, also includes providing challenging work, recognition, and personal growth. Other motivation factors could be job security, work culture, organizational policies, and supervisory practices. Organizations must understand that employees have a life beyond their offices, which is an equally important aspect of their lives. Keeping this in mind, organizations must make employee plans and policies to ensure that employees can facilitate a balanced life. Organizations should aim at providing their employees with a balance of monetary and non-monetary gains. They should also aim at creating happy and satisfied employees.

MOTIVATION: LEADING FROM THE FRONT

Shaheed Captain Vijayant Thapar (1977–1999) of the 2 Rajputana Rifles, Indian army was awarded the Indian military honour, Vir Chakra for his bravery during the Kargil War. Col. V. N. Thapar, father of Captain Vijyant Thapar had kept his last letter. The letter was written by Vijyant before he led the last assault that night as he knew that he will never come back. Vijyant always wanted to join the army and fight for the nation.[29] His letter made it crystal clear that the chances of his survival are remote. In such situations unlike most individuals he had no fear of pain or agony of losing his life or getting back to his family. Instead he was happy that he is going to sacrifice his life for his larger family; our country and hence, also mentioned in the letter that if born again he will join army and fight for the country. He sacrificed his life during Kargil war after valiantly fighting for about an hour and a half between fierce exchange of bullets and abuses. A burst of fire struck him on his head and he fell in the arms of his comrade Naik Tilak Singh.[30]

Such courage can only be exhibited by brave hearts with a lot of motivation. Such motivation drives them to accomplish goals that might even require sacrificing their lives. A soldier can possess such motivation only if he considers his profession as a voluntary action and has a sense of wanting for the nation. Traits such as motivation and leadership define Captain Vijayant. Even though the chances of victory were next to impossible, he showed immense courage and led from the front. His decision and determination to go to the forefront symbolize his leadership, which has motivated generations of Indian soldiers and the youth alike. His decisive sacrifice and outstanding bravery

(Box continued)

---

[29] Tarun, T. His Last Letter. http://www.captainvijyantthapar.com. [Online] http://www.captainvijyantthapar.com/lastletter.html (accessed on July 23, 2014)
[30] Tarun, T. The final assault. http://www.captainvijyantthapar.com. [Online] http://www.captainvijyantthapar.com/finalassault.html (accessed on July 23, 2014)

(Box continued)

> resulted in the victory of Kargil War. When the choice came between protecting his country and himself, he chose the former and laid his life for the honour of his country.
>
> Captain Thapar also mentioned in his letter, "As far as the unit is considered the new chaps should be told about this sacrifice." Because of the heroes like Captain Thapar, our country is safer today as they gave their today for our tomorrow. The letter he wrote to his parents before leaving for Knoll, voices volumes about his will to serve the motherland and shows his heroism. It is such heroism that binds the Indian armed forces and generates a resilient trust between the various ranks. Reading stories of such great heroes drives the youth to do something for our country and motivates them to join the Indian army.

5

ORGANIZATIONAL CLIMATE AND CULTURE: CREATING A CLIMATE OF TRUST

INTRODUCTION

In these turbulent economic times, it has become imperative for organizations to maintain a healthy and a positive work environment. Today, having efficient employees assures increasing productivity of an organization.

Corporate organizations are both similar and different from the military organizations. They have a similar organizational structure in terms of the hierarchy followed and the organizational dynamism. On the other hand, they are very different in terms of work enforcement, employment contracts, and organizational mission and purpose. However, the fact is that *the modern CEO functional staff is based on the German General Staff from the first half of the twentieth century.*[1]

Contemplating organizational ethics in military organization can throw light on the circumstances contributing to unethical behavior

---

[1] Traxler, R. N. (1961). A model of modern administrative organization: The German general staff. *Academy of Management Journal, 4*(2), 108–114.

of employees, corrective measures for improving unethical behavior, and methods organizations can use to influence individual and group decision-making processes.

The military follows a code of ethics; similarly, corporate firms also have a corporate code of ethics that the employees need to follow. Militaries across the world follow the best of organizational practices and ethics. There are many lessons corporate organizations can derive from them.

ETHICAL WORK CLIMATE

Work climate of an organization influences its employees' ethical decision-making ability, suitable to its mission and values. The US military is an apt example for this. It is the largest employer in the United States. About 1.1 million people are employed with the National Guard and Reserve Forces, and about 0.7 million people are employed in the civil personnel category.[2] There is no defined work climate that the military operates under, but it is definite that its work climate is extremely influential. The military's ethical work climate (EWC) is such that its culture influences the ability of its members to identify a problem, make the right decision, and understand how to act appropriately in any given circumstance.

Ethical Work Climate (EWC) has been defined by John Cullen as "the perceived prescriptions, proscriptions, and permissions regarding moral obligations in organizations."[3]

A study by Weber indicated that multiple sub-climates existed across different verticals in an organization. Therefore, different segments of an organization can have different cultures. He said that there were five types of an ethical work climate: (a) instrumental, (b) caring, (c) rules and procedures, (d) law and professional codes, and (e) independence.[4]

[2] United States Department of Defense. (n.d.). Retrieved from www.defense.gov/about/#mission (accessed on October 3, 2010).

[3] Cullen, J., & Bart, V. (1988). The organizational bases of ethical work climates. *Administrative Sciences Quarterly, 33*, 101–125.

[4] Weber, J. (1995). Influences upon organizational ethical sub-climates: A multi-departmental analysis of a single firm. *Organization Science, 6*, 509–523.

Different organizations require different work environments. In case of the military, where people are dealing with issues related to a nation's security, all the aforementioned elements of the work climate would be needed. An environment with *rules and procedures* and *law and professional codes* will assure discipline and dedication from the officers. However, military systems across the world allow certain amount of independence, and depict care toward their officers. This helps in keeping the officers going. The military has a unique work climate. The relationships and the bonding that colleagues in the military organizations share, go beyond mere professionalism. This is because they share a lot of good and bad things together, and, also, they understand each other better because of their similar situations. This similar trait has also been found in the corporate organizations. Studies have indicated that the work climate in an organization affects and determines the employees' attitude toward work, commitment to the organization, and the tendency of behaving unethically.[5]

Various studies indicated that the work culture of an organization influenced its employees' attitude toward their work and the employing organization.[6] From Weber's investigations, it was observed that an organization that followed an *instrumental* work culture reported more cases of theft, and in organizations that followed *caring* as their work climate, no cases of theft were reported. There is no perfect and predetermined work climate for any organization. The right work climate for an organization can be determined on the basis of workgroup characteristics and the organizational environment.

SHARED RESPONSIBILITY

Organizations must create the feeling of shared responsibility in their members, especially between members in different positions and verticals. In the military, the senior officers are in a position

---

[5] Trevino, L. K., Butterfield, K. D., & McCabe, D. L. (1998). The ethical context in organizations: Influences on employee attitudes and behaviors. *Business Ethics Quarterly, 8,* 447–476.

[6] Vardi, Y. (2001). The effects of organizational and ethical climates on misconduct at work. *Journal of Business Ethics, 29*(4/2), 325–337.

to voice their opinions on crucial issues, such as policy of weapons acquisition, and even on mundane operational issues. Leaders in the military are certain that their opinions and views are given serious consideration, irrespective of the senior-most supervisor's final decision. The subordinates in the military have to earn this. However, they are aware that their opinions and views do matter and are considered. The military believes in more than just giving orders. It believes that obtaining conformity and willful participation from the troops are also necessary.

Organizations should inculcate a communication pattern, wherein all the employees can contribute their opinions and viewpoints to improve the organizational systems. Organizations must treat all their members with equal importance. Organizations gain more knowledge and problem-solving solutions from the subordinates, if they respect their senior officers and get the same in return. Following the principle of shared responsibility, many changes in important policies in an organization can be easily implemented. Shared responsibility reduces conflict among members from different departments and verticals.

UNDERSTANDING AND RESPECTING ORGANIZATION'S CULTURE

An organization's culture indicates its values and mission. It is specially designed to create an intended environment conducive to the growth of the organization, employees, and stakeholders. To ensure that the organization's mission and employees' efforts are in alignment, it is important that the employees understand and value the work culture and climate.

Culture can be defined as "the prevailing values, philosophies, customs, traditions, and structure that over time have created shared individual expectations within an institution about appropriate attitudes, personal beliefs, and behavior."[7]

Studies have indicated that an organization's culture and its output are directly related. A positive work culture increases the

---

[7] Ulmer, Walter F., Collins, J. J., & Jacobs, T. O. (2000). American military culture in the twenty-first century. Washington, DC: Center for Strategic and International Studies.

output of each employee, thus increasing the productivity of the organization as a whole.

In case of an organization such as the military, the mission and the purpose of the organization are defined well in advance. Their purpose is to fight and win for the sake of national security. Those who enter the military understand and value the mission and the purpose of the organization. Great efforts are taken by the military to train its employees, as the duties and responsibilities bestowed upon them are very crucial. Members in the military are trained in a way that they are prepared to give up their lives in case the need arises. Not only are they physically molded, but also mentally strengthened to take the toughest decisions when on duty, and even in personal life. As Sarkesian stated, "the military profession stands and falls according to its ability to maintain and reinforce military culture."[8]

Although the military and the corporate firms have very different cultures, there are phenomena that corporate firms could adopt from the military. Decision making in corporate organizations is as difficult and stringent as in the military. It has been observed that central guidance in solving problems is found missing in many firms. In a lot of firms, inefficiency and low productivity arise out of lack of communication and direction from the supervisors. Corporate organizations need to work on this. They should keep their members at different hierarchal levels well communicated. Effective information sharing within an organization leads to quicker problem solving, and increased productivity of employees and the organization. When all the members know why certain changes are taking place in an organization, it is easy for them to adapt to those changes.

Military Culture

"The military profession stands and falls according to its ability to maintain and reinforce ... military culture."[9]

[8] Sarkesian, Sam C., & Connor, J. C. (2006). *The U.S. military profession into the twenty-first century: War, peace and politics* (2nd ed.). London: Routledge.
[9] Capstick, C. M. (2003). Defining the culture: The Canadian army in the 21st century. *Canadian Military Journal, 4*(1), 48.

Professor Donna Winslow has stated that culture

> represents the behavior patterns or style of an organization that members are automatically encouraged to follow. Culture shapes action by supplying some of the ultimate aims or values of an organization, and actors modify their behavior to achieve those ends. It establishes a set of ideal standards and expectations that members are supposed to follow. It is important to remember that culture is not only a set of values, or ethos, it is also the customary style used in organizing action.[10]

The military culture is a conglomerate of values, customs, traditions, and their philosophical foundations that, over time, has created a shared institutional ideology. The military culture brings a common framework for men in uniform, and a common standard of behavior, discipline, teamwork, loyalty, selfless duty, and the customs that support those elements.[11]

Separateness of the Military from Other Professions

In 1869, William Windham described armed forces, generally, as "a class of men set apart from the general mass of the community, trained to particular uses, formed to peculiar notions, governed by peculiar laws, marked by peculiar distinctions." As a result of its uniqueness and the need to instill organizational loyalty and obedience, most military organizations develop a culture unto themselves, distinguished by an emphasis on the hierarchy, tradition, rituals and customs, and distinctive dress and insignias.[12] The notion of unlimited liability in defense for national interests makes a big distinction between members of the military profession and other professions. Moreover, a soldier sacrifices his personal life in order to achieve military objectives.

---

[10] Ibid., 48.
[11] Ibid.
[12] Anonymous. (1997). Military Culture and Ethics. *Report of the Somalia commission of inquiry, 1*. Retrieved from http://www.dnd.ca/somalia/vol1/v1c5e.htm (accessed on July 31, 2014).

Regimental Culture

In the British and Canadian armies,

> the regiment is an extended family that reaches backwards in time and outwards in space to encompass those soldiers who have come to identify with its collective memories and traditions. Each regiment develops a culture that is partly rooted in the place from which it draws its members and partly in a set of values and mores that have been created for the sole purpose of making it different from other regiments.... For the most part, their [the soldiers'] life and loyalty centre on the regiment—not on the army.[13]

The military culture of a nation is subdivided into various military regiments. These territorial divisions (regiments) define areas of recruitment, training, and residence for regimental members. This regimental subculture provides a common bond, uniting its members. According to Major Gen (Retd.) Dan Loomis, regiment is a pseudo-kinship organization. It is often referred to as a family, and according to another analyst, its essence is tribal and corporate, rather than instrumental and bureaucratic. One is considered a member of a regiment for life. This link continues throughout a member's career in the military and after retirement. According to Major Gen Loomis, "The Regimental Family permeates all facets of one's life from pseudo-birth as a new member to death."[14]

It is a well-accepted fact that a soldier's regiment is his family, and in a battlefield, a soldier risks his life for his comrades, and for the honor and survival of his regiment. Many officers and soldiers contribute their entire lives to a single regiment, and they naturally become blind to many of its faults.

James Burk, a prominent academic observer of the American military, has developed a new model of military culture which consists of four essential elements: discipline, professional ethos, ceremony

---

[13] Capstick, C. M. (2003). Defining the culture: The Canadian army in the 21st century. *Canadian Military Journal, 4*(1), 49.

[14] Anonymous. (1997). Military Culture and Ethics. *Report of the Somalia commission of inquiry, 1.* Retrieved from http://archive.today/CMkd

and etiquette, and cohesion and *esprit de corps*. These are workable and present in any military culture or subculture. The requirement for a "clear vision of the desired institutional (military) culture" is unequivocal. It must be accompanied by a strong resolve by senior leadership to ensure that the vision is realized and that it becomes the essence of military service.[15]

CHOOSING GOOD PEOPLE

Corporate organizations should choose employees whose personal goals and missions align with the organization's own goals and missions. In case of the military, personnel selection standards are continuously upgraded and revised. Personnel chosen for different tasks undergo rigorous training and difficult tests. The most favorable factor is that the prospective employees know what they want to achieve by joining the military and why they want to do it. Their initial training phase is difficult, but they choose to go through it. The brutal and harsh training prepares the officers to be stronger for the tougher times to come. Their jobs are very demanding. However, the military manages to derive the best out of them. Even though they are yelled at and punished severely for their smallest mistakes, it does not make them feel any less than the others in the league. The punishments and reprimands they receive are meant to train the officers to follow the military system of commands. Once they have joined and dedicated themselves to the military, their world changes. They are willing to make a lifetime commitment to their duties and take it forward to the next generation.

Corporate organizations could adopt selection standards used for selecting military personnel. These standards should be competitive and high. The selection standards should also be upgraded and revised with time. There is a vast difference between the way people in a corporate and those in the military would be treated, but the essence remains the same. Corporate organizations should treat all the employees with equal respect and dignity. In case of a

---

[15] Capstick, C. M. (2003). Defining the culture: The Canadian army in the 21st century. *Canadian Military Journal*, 4(1), 49.

fatal error, they should be provided with counseling and required training for the first time of an error, but later, stern warnings and penalties should be implemented for repeated errors. Organizations should focus on their core values, and the same vision should be inculcated in the employees as well. A clear and defined set of core values assures stability, continuity, and increasing efficiency of a firm. Leadership development must be given high priority, as today's associates are going to be tomorrow's managers. Their skills must be assessed on a regular basis, and their feedbacks must be collected at regular intervals. Employees should also be provided with continuous required training to improve their skills, and, thereby, match the global demands. In the army, training cannot happen in a day's time, or even in a short span. Consistent and conscious efforts should be made to master the skills. Simultaneously, other skills such as soft skills, social skills, etc. should also be inculcated in them.

TAKING CARE OF/CONSIDERING HUMAN NEEDS

Maslow explained human needs in five levels. He called it the *hierarchy of needs*. He said that as human beings accomplish one level, they aim for the next higher levels. In the first level, they try to fulfill the basic necessities (food, clothing, and shelter), and then move to the second level to achieve security, stability, and protection. Moving to the third level, they strive to belong to a group and be accepted. In the fourth level, they endeavor to satisfy the need of self-esteem. In this level, they look for appreciation, which boosts their self-esteem and self-confidence. Moving to the fifth level, they try to cater to their self-actualization needs. Here, they develop their career goals and plans and develop skills, through training to achieve the goals, which lead them to self-development.

As per FIRO-B survey, the participants reported their needs in the form of inclusion, affection, and control. It was observed that the participants expressed their desire to have honest and open relationships. They assigned highest priority to honest relationships, communications, and greater bonding. All leaders must understand the needs of their employees and provide them with whatever they can, so that their increased productivity and loyalty are assured.

If the leader assures that the subordinates' needs, such as recognition, appreciation, respect, and approval, etc., are fulfilled, the probability of an organization achieving its mission and vision increases.[16]

ORGANIZATIONAL CITIZENSHIP BEHAVIOR: RELATIONSHIP BETWEEN SUPERVISORS AND SUBORDINATES

Relationship between the managers and the subordinates is very important to increase the productivity of an organization. This is called the organizational citizenship behavior (OCB). OCB is majorly influenced by the procedural fairness in a firm. Procedural fairness can be explained as the means by which organizations and their representatives make allocation of decisions. Procedural fairness can be experienced by the subordinates when supervisors behave in accordance with the OCB. Procedural justice builds trust for the organization in the minds of the employees, and also makes them content about their work.[17] It is important that the supervisors are content with their jobs. Studies have indicated that supervisors' satisfaction is related to the degree in which they show consideration and supervisory mentoring behaviors. Supervisory mentoring behaviors generate more favorable procedural justice. Green and Bauer stated in their study that supervisory mentoring *leads to an increase in supervisor's salary, productivity, and promotability.*[18]

DEALING WITH DYSFUNCTIONAL BEHAVIOR IN ORGANIZATIONS

Many organizations face the problem of dysfunctional behavior in employees. Such behavior hampers an organization's productivity, and, thus, affects its profits. In case of the military, officers' behavior

[16] Klann G. (July/August 2001). Leading a changing workforce lessons from the U.S. army. *Leadership in Action, 21*(3), 8–11.

[17] Tepper, J. B., & Taylor, C. E. (February 2003). Relationships among supervisors' and subordinates' procedural justice perceptions and organizational citizenship behaviours. *The Academy of Management Journal, 46*(1), 97–105. Retrieved from http://www.aom.pace.edu/amj/february2003/tepper.pdf (accessed on July 31, 2014).

[18] Green, S., & Bauer, T. N. (1995). Supervisory mentoring by advisers: Relationships with Ph.D. student potential, productivity, and commitment. *Personnel Psychology, 48,* 537–561.

is monitored strictly, as they are exposed to dynamic battlefields and warfare, affecting the national security. Dealing with situations like these requires strict discipline, skills, and efficient decision making. There is no scope for dysfunctional behavior in the military.

A solution to dysfunctional behavior at work is inculcating positive group context. Positive group context is an essential element of an ethical work environment in any form of an organization. A positive group context is defined as a situation in which group members perceive the presence of encouraging work conditions that enhance their attitudes about the workplace; these can include helping behavior, group cohesiveness, peer leadership, and cooperation in completing work requirements.[19]

Lack of positive group context could lead to misbehavior in the form of *withholding of effort*. Better management of group-level tasks and effort-to-performance linkages could increase the productivity of the employees as well as of the organizations. Withholding of effort can be defined as "the likelihood that an individual will give less than full effort on a job related task."[20] There could be several reasons for this. One reason could be holding back from making full effort to the task assigned, neglecting the work assigned, and reduced efforts with others around. Neglecting work and lack of interest could be an outcome of the lack of required training, guidance, and communication. Withholding of effort not only leads to low productivity of the employee, but also has a negative impact on other employees, reducing the overall productivity of the organization. If withholding of effort of an employee is ignored by the authorized manager, other employees who are taking efforts to maintain harmony at the workplace become demotivated because of tolerance of such conduct. When organizational climates are perceived as being more supportive socially and emotionally, they generally tend to lower the level of organizational misbehavior.[21]

[19] Vardi, Y. & Weitz, E. (2004). *Misbehavior in organizations*. Mahwah, NJ: Lawrence Erlbaum Associates.
[20] Kidwell, R. E. Jr., & Bennett, N. (1993). Employee propensity to withhold effort: A conceptual model to intersect three avenues of research. *Academy of Management Review, 18*, 429–456. doi:10.2307/258904.
[21] Ibid.

DEALING WITH STRESS

Employees in the military are exposed to tasks that can leave them traumatized, depressed, sad, and demotivated. There are various kinds of stress that the people in the military face. Lack of affection, home sickness, lack of freedom, work pressure, competition, and adverse living conditions are the common stress factors. The military makes arrangements for its stressed employees. Timely counseling, entertainment, communication with family, etc. are facilitated to comfort the members and to take care of them.

Stress among the working class people has become a common phenomenon. With the increasing work pressure, deadlines, and extra working hours, employees are often found tired and depressed. This pressure only increases as an individual moves up in the career path. The current lifestyle, where people work till late hours, have sedentary jobs with no time for physical exercises, and spend less time with family and friends, has contributed to the increased stress in many employees.

- **Boosting Morale and Homecoming Experience**
  Group cohesion and support are important factors that determine the morale of the employees in an organization. Lubac explained that high morale leads to less stress among employees.[22] Organizations should boost the morale of the employees through continuous training and morale-boosting sessions. Conducting team-building activities has proven to be a good morale-boosting tool. When members of the organization from different departments and verticals come together and indulge in interactive group sessions, the positive and friendly work environment, and group cohesion boost their morale.
- **Dealing with Stigma**
  Stigma can be explained as a sign of disgrace or discredit that sets a person apart from others.[23] There are various factors that

---

[22] Labuc, S. (1991). Cultural and societal factors in military organizations. In R. Gal & D. Mangelsdorff (eds), *The handbook of military psychology* (pp. 471–489). New York: Wiley.

[23] Byrne, P. (2000). Stigma of mental illness and ways of diminishing it. *Advances in Psychiatric Treatment, 6,* 65–72.

can cause stigma. In case of the military, if an officer makes a wrong decision that leads to loss of some other officers, the guilt and the stigma for this will haunt him. Similarly, in a corporate organization, supervisors and subordinates have to often make important decisions. One wrong decision affects the whole team, department, and the organization. There should be ways through which organizations try helping their efficient and promising employees to get out of a stigma, as this could cause mental stress, lowering the individual's productivity. Buddy system opted by the military, where individuals are taught to rely on each other, could be adopted by the corporate organizations.

- **Dealing with Social Issues in an Organization**
  Military is one of the institutions that has not only dealt with, but also brought about solutions of social issues such as gender discrimination, and partial and biased gender preferences. Instead of continuing with the conventional beliefs of women's physical limitations, the military has now come up with improvised training programs to bring women at par with men as combat frontlines. The military has set a unique example for bringing change in the social issues of gender discrimination. Trainings of men and women were initially conducted only for the air force, but now, an integrated training for the army, navy, and air force has been initiated. In the past, homosexuality in the military was not tolerated. However, with time, as homosexuality became a norm, the military has become more tolerant toward it. The military has been revising and updating its policies to accommodate the emerging and already existing social issues.

Corporate organizations should be progressive too. With globalization, various issues and social complexities can arise that did not exist before. Companies should have policies in place that can make people liable for committing any social discrimination.

- **Less Women at the Top**
  From the historic times, fewer women have attained the top positions in an organization. Even today, after implementation of

gender-neutral promotion systems, not many women are at the top positions. A general view has been that women themselves choose to not attain higher positions in an organization, because of their families and the commitment to bring up their children. The military, in this aspect, has facilitated women with a lot of comfort and freedom. They are provided with support that can ease their family pressures and commitments. This gives them the freedom to work passionately, and, simultaneously, pay attention to domestic matters. "Women in the military account for only 2% of the flag officers, 6% of the senior officers, 8.8% of the junior officers and 11.4% of the noncommissioned officers."[24]

Although many organizations follow gender-neutral systems of tests, not many women are still at the top managerial positions. Women in corporate organizations account for only 17 percent of the total top management positions.[25] Organizations should try and promote women to work by providing them with help such as crèches, help at home, relocation of spouse, flexible work timings, etc.

- **Misbehavior at Work**
  There are incidents of misbehavior by employees in several organizations across the globe. This could be in the form of assaulting women, breach of data security law, racism, etc. The military has zero tolerance for any such misbehavior. The officer in question, if proven guilty, is suspended, ousted, or even forced to resign. The military is known for its discipline and the gentlemen it creates. The onus of this lies with the discipline, and the rules and policies followed in the military to maintain the harmony and integrity within the institute.

  It has been found that despite the rules and polices implemented to control and monitor any sort of misbehavior,

[24] NATO Committee on Gender Perspectives. *National Report from Norway 2009*. Retrieved from http://brage.bibsys.no/xmlui/bitstream/handle/11250/99263/01%20Gender%20and%20Military.pdf?sequence=1

[25] Martinson, J. (January 30, 2012). Why are women stuck at 17% of top jobs? [Web blob post]. Retrieved from http://www.guardian.co.uk/lifeandstyle/the-womens-blog-with-jane-martinson/2012/jan/30/few-women-in-top-jobs (accessed on July 31, 2014).

corporate organizations have been lenient toward certain forms of behavior or biased toward certain members of the organization.[26] This is detrimental to the reputation of the organization in front of the world, and even its own employees. A company's values create respect in others for it. If this fails, chances of higher rate of misbehavior and situations of social unrest can arise. This will, in turn, have a direct negative effect on the productivity of the firm. Organizations should take strict actions against employees who have misbehaved, so that others also can learn a lesson from it. A secure and supportive work climate will induce employee trust and loyalty.

EXERCISES THAT CAN STRENGTHEN THE ORGANIZATIONAL CULTURE

The military takes up many exercises to strengthen its organizational culture. The military's culture lays the foundation of its values and mission for its members (Figure 5.1). Let us look at them one by one.

- **Workplace Design**
  An organization's workplace design indicates its strategic environment. Many organizations have their offices scattered in different cities and countries. It has been observed that scattered workplaces affect the social-learning constructs. This was in the form of reduced morale, inefficient information sharing, and lack of communication among the members of the organization. People preferred face-to-face communication to solve problems and gain guidance. It was also found that people employed in smaller outposts often belonged to junior ranks. They felt very little identification with the organization they were working for. Employees located in different offices were sensed with confusion about the work culture and common identity. They considered themselves

---

[26] Kidwell, E. R., & Valentine, S. R. (2009). Positive group context, work attitudes, and organizational misbehavior: The case of withholding job effort. *Journal of Business Ethics, 86,* 15–28. doi:10.1007/s10551-008-9790-4.

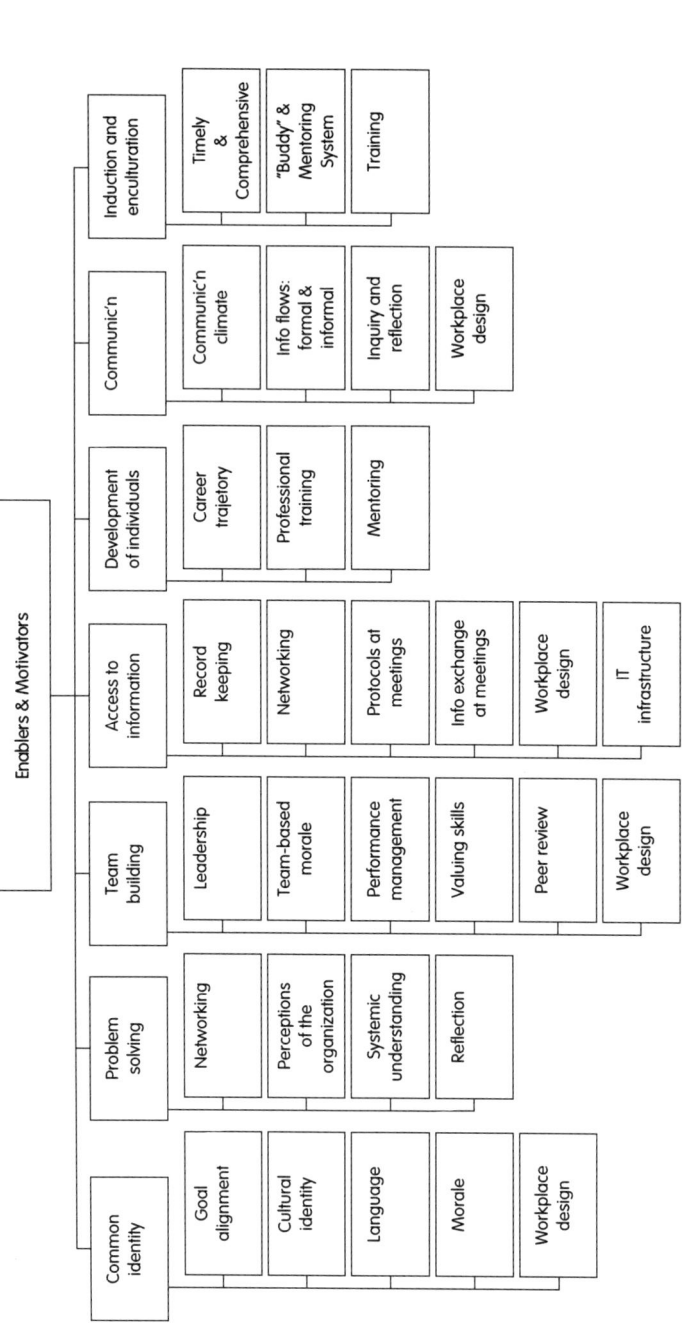

Figure 5.1 Social-Learning Constructs

Source: Warne, L., Ali, I., Pascoe, C., & Agostino, K. (December 2001). A Holistic Approach to Knowledge Management and Social Learning: Lessons Learnt from Military Headquarters. *AJIS Special Edition Knowledge Management, 9*(1), 127–142.

as the outliers in the organization. An opposite opinion was received from those who worked in the headquarters. Efforts should be made to keep all the employees of an organization well connected and communicated. It is also very important that all the employees are treated equally. A workplace located far-off should be designed in such a way that its environment reflects the organization's values and culture. This will increase the efficiency of the employees, and, thereby, increase the efficiency of the organization.

- **Networking**
  Personal and professional networking is an important tool for knowledge sharing. Knowledge is transferred from one person to another. This also leads to an increase in the productivity of those who have string networking and quick exchange of information. Networking is conducive to an individual's career growth as well as personal growth.
- **Team-Building Exercises**
  Team-building exercises are proven as efficient ways to get the team members connected. Team-building activities help in building healthy professional relationships among the members and in building a bond between them. When the employees in an organization are friendly and warm toward each other, their efficiency at work is improved. Working in such an environment induces happiness in the employees and makes them work efficiently. Workplace then becomes a place where people willingly come to work, and they enjoy what they do.[27] Team-building activities build team-cohesiveness and positive team spirit. Employees working in teams derive better results than those working by themselves.
- **Performance Management**
  Performance management is an important part of the overall management of the military personnel. Performance management calculates individual performance progression. This helps

[27] Warne, L., Ali, I., Pascoe, C., & Agostino, K. (December 2001). A holistic approach to knowledge management and social learning: Lessons learnt from military headquarters. *Australasian Journal of Information Systems* (Special Edition Knowledge Management), *9*(1), 127–142.

employees in understanding their weak areas and improving on them. It also gives the companies a forecast of the best performers in the organization to appoint future managers and plan employee promotions. Similarly, team-based performance management systems can be implemented. This will help the firm in understanding the efficiency of all the departments individually and collectively.

- **Peer Review**
  In the military, peer reviewing is an important way of sharing individual expertise and problem-solving skills. It is also an effective team-building practice. Peer reviewing showcases an employee's willingness to help and assist his/her fellow colleagues. It is a good exercise for building sustainable professional relationships on the basis of trust. Peer reviewing is an efficient way of receiving feedback, criticism, and substantiation. Peer reviewing facilitates knowledge sharing and social learning among the members.

- **Effective Information Exchange and Communication**
  Information sharing and exchange is an important organizational resource that facilitates increased productivity and improved decision making. Access to information should be easy for the employees. Easy access to information acts as a direct input for knowledge acquisition and generation, and also social learning. Furthermore, the Internet has facilitated smooth flow, and easy access and transfer of information across the world. Information about amendments made in meetings and other related information should be passed on to the employees. This can bring opportunities for brain-storming discussions and group reflection.

- **Training and Development of Individual Expertise**
  Timely professional training is an important tool for the development of individual expertise. Organizations should pick efficient employees and send them for further training and skill-upgradation sessions. This ensures that the organization is creating efficient and well-equipped leaders for tomorrow. Furthermore, this also creates a feeling of trust and loyalty in the employees' minds, motivates them to perform better each time, and assures them a career growth.

- **Mentoring and Motivation**
  Mentoring of juniors by supervisors is an important tool of assisting the development of junior staff. Knowledge received from mentoring can go a long for the mentee. This is because mentoring involves personalized attention and teaching lessons learnt from experience, and not just imparting theoretical knowledge. Mentoring can also prove to be beneficial to the juniors to determine the right career path.
- **Efficient Communication**
  Efficient communication is the lifeblood of an organization. Efficient communication systems are a source of effective learning and knowledge management in an organization. A supportive communication climate should be created in an organization. Such a climate leads to an effective, free, and open exchange of information, and a high rate of employee involvement in problem solving. A supportive communication climate induces a culture in which all the participants are treated equally and with respect. This can help in breaking down the cultural barriers that lead to inefficient communication. Studies have indicated a link between supportive communication in an organization and higher level of commitment with the organization.
  The communication systems in organizations should promote and facilitate an effective formal and informal exchange of information. This exchange of information helps in aligning individuals' goals with the goals of different departments and verticals in the organization. Formal flow of information would indicate sharing of information from the top level to the bottom level and vice versa. On the other hand, informal flow of information would indicate a work climate conducive to informal chats and discussions among groups. This is a form of socializing, team building, and sharing of information and knowledge. The work climate should be such that besides working, employees should get the opportunity to interact with their colleagues on a personal and informal level.
- **Induction and Enculturation**
  Induction and enculturation sessions for new employees are often the ignored and overlooked aspects of new-employee

training. An elaborate and informative induction and enculturation session for the new employees assures instant understanding of the firm's values and mission. It helps the employees connect and establish a bond with the organization they are working for. Induction sessions help the employees in attaining knowledge about the processes in the firm and the industry, and facilitate social learning. It helps the employees understand the career growth prospects in the firm, and ensures that they have just expectations. It is an efficient way of building relationships, meeting new people, building morale, and understanding rules and regulations.

Induction programs build the base for employees about what kind of work they are going to do, how to do it, and whom to consult for guidance and help.

ORGANIZATIONAL CULTURE: DEVELOPING CAMARADERIE

In order to promote the friendly spirit that should exist in every unit, the custom of "calling on" is a duty that every officer is expected to observe. It is one of the principal means whereby all personnel, including the families of officers, get to know each other. Calling on also helps in creation of rapport and enables new officers to get attuned to the working culture of the organization.[28]

Calling on is of two types:

- The first one is the official call. This call is to be made by the officer to his immediate superior at the first immediate opportunity. It is customary to obtain a convenient time to make the call. Official calls are made in full ceremonial uniform and should last for fifteen minutes.

(Box continued)

---

[28] Til wheels are up. Dining In/Out http://www.militarywives.com. [Online] http://www.militarywives.com/index.php/protocol-mainmenu-264/air-force-protocol-mainmenu-298/dining-in—out-mainmenu-342 (accessed on July 23, 2014).

(Box continued)

> - Second one is social call. Usually, the officer who arrives last in the station calls on the others already present. And these calls are generally made along with wives and should last for 30–45 minutes.[29]
>
> The purpose of calling on is to bring together a unit in an atmosphere of camaraderie, good fellowship, and social rapport. The basic idea is to enjoy oneself and the company. Calling on is also an excellent means of bidding farewell to the departing members and welcoming newly arrived members to a unit. It is an excellent forum to recognize individual and unit achievements. Calling on, therefore, is very effective in building high morale and esprit de corps.[30]

---

[29] Chapter 4 Mess etiquette and customs. http://www.irfc-nausna.nic.in. [Online] http://www.irfc-nausena.nic.in/irfc/ezine/etiquette/chapter4.htm (accessed on July 23, 2014).

[30] Chapter 2 Military courtesies. http://www.irfc-nausena.nic.in. [Online] http://www.irfc-nausena.nic.in/irfc/ezine/etiquette/chapter2.htm (accessed on July 23, 2014).

6

DEVELOPING SOPs: STRATEGIES AND TACTICS

INTRODUCTION

In this global market endowed with cut-throat competition, firms need to implement cutting-edge strategies to compete with other firms. Firms need to be aggressive, vigilant, quick, well informed, and well prepared with everything that gets them to the top. Military as an organization is the best example of this.

There are various lessons that the corporate firms can learn from the operational procedures followed by the military (Figure 6.1). Military standard operational procedures (SOPs) consist of various strategies and tactics to compete with the rivals. One of the most crucial lessons that corporate firms can learn from the military is identifying the right strategies and executing them at the right time.

USING THE RIGHT STRATEGY

In the past, many critical wars have been won by the sheer skill of using the right strategies. A good depiction of this is the Athenian versus Persian war, which occurred in 480 BC. In this war, the Persian ships were more in number than the Athenians. Keeping

DEVELOPING SOPs: STRATEGIES AND TACTICS

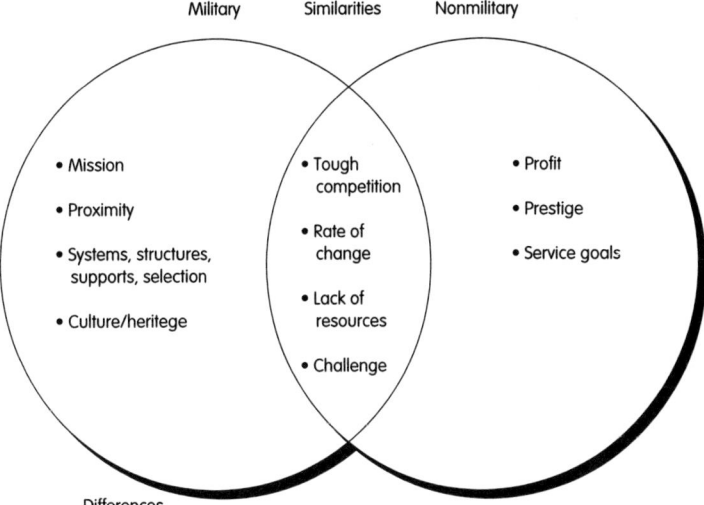

Figure 6.1 Similarities and Differences between the Military and Corporate Firms
Source: Author's own.

this in view, Athenian General Themistocles lured the Persians to move toward the Straits of Salamis. He was aware of the superiority of the firepower of the Greek warships and used it to his advantage. Themistocles started attacking the Persian ships from a point where only a few ships could fight at a time. Themistocles took advantage of the superior firepower of the warships and defeated the Persians.[1]

The aforementioned historic example teaches the corporate organizations to use the right strategy to beat the competitors. The crux of using the right strategy lies in taking the right step, at the right time, and at the right place. This is the best way to put your competitor at a great disadvantage.

The term *strategy* has been defined in multiple ways by various authors, but almost everybody followed a common platform, that of a calculated conscious set of guidelines that determines decisions into the future. In game theory, strategy represents the set of rules that is to govern the moves of the players. In military theory, strategy

[1] Widmer, H. (1980). Business lessons from military strategy. *The McKinsey Quarterly, 2,* 65.

is "the utilization, during both peace and war, of all of the nation's forces, through large-scale, long-range planning and development, to ensure security and victory" and in management theory, the Chandler definition is typical: ". . . the determination of the basic long-term goals and objectives of an enterprise and the adoption of courses of action and the allocation of resources necessary for carrying out these goals." All these definitions treat strategy as (a) explicit, (b) developed consciously and purposefully, and (c) made in advance of the specific decisions to which it applies.[2]

The formation of strategies in most of the organizations revolves around three basic forces:

- A continuously changing environment with irregularities, discontinuities, and wide swings in its rate of change
- An organizational operating system or bureaucracy that aims to stabilize its actions, despite the characteristics of the environment it serves
- A leadership, whose role is to interlink the other two forces and to maintain the stability of organization's operating system by insuring its adaptation to environmental change

Strategy can then be viewed as the set of consistent behaviors by which the organization establishes for a time its place in its environment and strategic change can be viewed as the organization's response to environmental change, constrained by the momentum of the bureaucracy and accelerated or dampened by the leadership.[3]

There are three types of strategies laid down by Widmer in his study *Lessons from Leadership Strategy* that corporate firms can use from the military.[4]

- **Offensive Strategy**
  Offensive strategy is usually designed in a way that the team reaches its objective irrespective of its enemy's attack.

---

[2] Mintzberg, H. (1978). Patterns in strategy formation. *Management Science,* 24(9), 935.

[3] Ibid., 941.

[4] Widmer, H. (1980). *McKinsey Quarterly,* 2, 65.

DEVELOPING SOPs: STRATEGIES AND TACTICS

This strategy can be segmented in four groups.

  a. **Frontal Assaults**: This would mean attacking the enemy head-on with all the required resources to attain victory. In business terms, this would indicate that all the firms should be well prepared with the course-of-plan. Firms should be ready with new products, new marketing strategies, and innovative and creative models to make their product standout in the market, thus attacking other products in the market. This is an aggressive strategy and is not considered as a viable strategy in the corporate world.
  b. **Dispersed Attack**: This strategy suggests attacking the enemy more subtly. This is done by attacking the enemy on its weak points one by one. In business terms, this can be explained as competing with the rivals by entering one market segment at a time. Using this strategy, firms can compete with their rivals on the basis of changes in the market forces which are not in either party's control.
  c. **Flanking Maneuvers**: This strategy involves disorienting the enemy line by applying pressure from both the sides. Here, the attacker takes advantage of the chaotic conditions caused by the two-sided pressure on the enemy line, and attacks the enemy in opportune. In business terms, direct advertisements and public relations can be a form of this strategy.

- **Defensive Strategy**

  Defensive strategy aims at protecting one's own interest. This strategy has four rules that need to be followed.

  a. Every attack must be made with the same amount of intensity.
  b. All important assets must be protected.
  c. One must always accurately evaluate the strength of the attacker.
  d. The best defense resides in the courage to attack your own positions.

  There are two types of defense strategies.

  a. **Fixed Defense**: This indicates munition of the defendable location. The limitation of this strategy is that the defender,

in this case, is highly immobile and his position is easily predictable. In case of business firms, this strategy can be implemented by focusing on the marketing strategies, such as provision of quick customer service and implementation of schemes that lead to high customer loyalty.
   b. **Mobile Defense**: This strategy involves constant changes in movement from one place to the other and in the tactics used. In this strategy, the enemy finds it difficult to trace the attacker. In corporate terms, this strategy can be used by using product differentiation, changing the positioning of the products and advertisements.

- **Guerrilla Strategies**
Guerrilla strategy uses the tactic of weakening the enemy by attacking it several times, in small breaks and in small intensities. In this strategy, the task force is divided into small segments, which are guided to attack the weak sides of the enemy in a selective manner. In business terms, firms can use this strategy by advertising their products in a way that it appears to be superior to other products in the market. This can be done by making reductions in the price of their goods, and also by using the tool of publicity.

Sun Tzu in his famous treatise, *The Art of War*, stated that "He who would avoid what is strong must strike at what is weak."[5] This statement indicates that you must not focus on the strength of your competitor, instead assess his weak points and attack him at his weakest point. This statement is a mantra that organizations should follow to beat their competitors. Corporate firms must understand the strength of their products or services, and the weaknesses of their rivals' goods and services. They should try to provide superior goods at the right price to the customers to achieve an absolute advantage over other firms. To attain this, performances of every component in the business process must be measured. Components with higher cost must be replaced with suitable cheaper options,

---

[5] Tzu, S. (2009). *The art of war: The oldest military treatise in the world*. Part 6. (Lionel Giles, Trans.). New Delhi: Radha Press.

alongside maintaining the quality. Investments should be made to improve the technical and engineering know-how.

DECISIONS BASED ON KNOWLEDGE AND NOT ASSUMPTIONS: KEEP YOUR FRIENDS CLOSE AND ENEMIES CLOSER

A common observation is that many firms' decisions made by the strategists are based on assumptions and not based on what actually exists. Strategists need to constantly evaluate and consider the movements within the organizations and market conditions, and accordingly make apposite decisions. With globalization, various business phenomena are experiencing quicker changes. Firms will need to keep these changes in consideration and make aligning changes in business decisions.

Consider a hypothetical example of an oligopoly market with three firms, Firm A, Firm B, and Firm C. These firms are involved in manufacturing and supplying homogenous goods at ₹1,000 per unit, and have one-third market share each. Now, to obtain a greater market share, Firm A reduces the price of its product to ₹900. As a counter act, Firm B reduces the price of its product to ₹850, which is the minimum price for firms A and B can quote to derive minimum profits. Following the trend, Firm C reduces the price of the product to ₹800. However, firm C could reduce the price of its product below ₹850, as it had a cost advantage over firms A and B, because it used advanced machinery, which produced more goods at a lower price. Firm C could, thus, capture the majority of the market demand for that product. Firm C in this case was aware of the fact that other firms could not reduce their prices further down. It, thus, made a smart move on the basis of the knowledge it had about its competitors' cost of production. Firm C enjoyed a win-win situation in either way.

Considering the same example, it can be said that Firm A's action of reducing the price of its product was not wise. Firm A, before lowering the prices had one-third share in the market. However, after reducing its price to ₹900 and inviting price war, it was left with a negligible market share. Firm A, in this case, should have first gathered more knowledge about other firms in competition, instead of assuming their cost of production.

SETTING THE TARGET

The military and the business firms do have common analogies. Militaries have enemies, they aim for terrains, and own troops, all to accomplish the mission of national security. Similarly, firms have competitors, they aim for customers, and own companies, or all of these, to achieve the business mission of making profit. Targeting in military is a factor that can determine the success or the failure of its mission. This scenario is not very different in case of business firms, although its perception is. When it comes to targeting, corporate firms limit themselves to adopting objective-setting, unlike the military. There is more that can be learnt from the military. The general perception is that firms are meant to survive in the market with many competitors. Their aim is to survive the competition and make sufficient profits. However, the reality is that the military terms such as "capture" and "occupy" do hold prominence in the corporate firms. Even though there are several competitors in the business world, in many cases, eventually, there is just one winner, who has an edge over other firms. Corporate firms find themselves in similar situations a lot of times.

For example, in the telecom industry, telecom providers with the highest bidding are given the contract. In this case, the target is not only for winning the bid, but also for attaining all the other associated expansions and experiences with this victory. A firm's victory in an industry indicates its stay in the market for a longer period. When two firms with the same capacities, in every aspect, are competing with each other for a single bid, there is just one winner. When the price and quality of goods/services provided by the two firms are same, the company that has a higher cost advantage will be able to accommodate add-on features or facilities, such as automation of services and charging the same price. This will, thus, enable the firm in gaining a comparative advantage over the competing firm.

Widmer argued that the aforementioned situation will not hold true for firms of all sorts. In case of banking firms, steel manufacturing firms, and watch manufacturing firms, there is never a single winner. In banking, almost all the parties are involved in making profit; in case of steel, no one really makes profit; and in case of watches, the newer firms derive more profits than the older ones. Widmer explained these situations using three circumstances. He said either one of the circumstances, as mentioned, is responsible for the prevention of a single winner.[6]

- There is an existence of cartel
- Mistakes have been detected from the market leader's end
- Absence of direct competition

Widmer explained that in case of absence of direct competition, a firm does not lose or win in a single market, but within a niche segment in the market. He explained this on the basis of a superficial observation. He indicated that firms with the ranks five, nine, or seventeen in the market are all operating with a profit, even though there are firms at the top. He explains this with the existence of marker segments. Firms with rank number one, two, or seventeen have a share in the market in the form of market segments.

Military teaches the corporate firms to define their target and choose the area in which they want to compete or can have an edge over other firms. Once a firm makes this choice, it will grow to be a clear winner.

CONCENTRATING ON THE COMPETITOR'S WEAKNESS

Concentrating on the enemies' weaknesses is one of the prime strategies that the military follows. Sun Tzu in his book, *The Art of War*, states that the direct way to attack the enemy is by directly joining the battle. However, he suggests that to secure one's victory, one must use an indirect method of attacking and refrain from directly attacking the enemy.[7]

The aforementioned Tzu's principle may sound irrelevant in the business world, but this can, however, be applied. Tzu's principle can be applied by corporate firms which cater to a niche market.

---

[6] Widmer, H. (1980). Business lessons from military strategy. *McKinsey Quarterly, 2,* 65.
[7] Tzu, S. (2009). *The art of war: The oldest military treatise in the world.* Part 6. (Lionel Giles, Trans.). New Delhi: Radha Press.

According to the general business trend, a firm will try to sell its goods in all the market segments to maximize its sales and consequent profits. However, this will not hold beneficial for all firms and business models.

For example, suppose Firm A manufactured high-end tractors and some other small agricultural machinery. It launched its tractor in the market. The tractor proved to be a time saver. It consumed about 30 percent less time in accomplishing the task compared to all its competitors in the market. Despite the superiority of the tractor, Firm A realized that it was not able to sell the tractors in all the market segments. Only the big farmers bought it, which was approximately 20 percent of the total market segment. Firm A realized that while small farmers could not afford the tractor, certain others were not even looking for a tractor that could save time.

In this case, Firm A can either focus on catering to the 20 percent of the niche market that holds the demand for its current products/services, or it can redesign and recreate products/services that would cater to the remaining 80 percent of the market. However, in the above case, Firm A must stick to providing to the niche market segment because it has a competitive edge over others. Among the other firms that manufactured the same products/services, Firm A provided the best model with a unique design at the same price. Firm A, by concentrating on the market segment it had a demand in, could dominate the market segment, and thus achieve greater profits.

It is a general observation that firms do not provide/manufacture on the basis of market segmentation. This indicates lack of concentration of resources in the market segments, where firms can derive maximum outputs. Market segmentation can be made on the basis of age groups, income levels, culture, etc. Without concentration of inputs, firms cannot make efficient strategies or choose the right target. Identifying and defining the focus market segment will enable applying the right strategies, at the right time, and at the right place.

The problem of lack of concentration can also be applied to the various policies implemented by the government. Majority of the government policies are applicable to all the citizens. This would lead to giving more benefits to the higher income groups or the better-off class, than the lower-income groups.

For example, the government imposes a fixed tax rate on a new medicine for malaria. People from the higher income groups will be able to easily afford the medicine. Whereas, people from below poverty line (BPL) will not be able to afford the medicine that could result in their death. This indicates short-sightedness of the government.

EMPLOYEE MANAGEMENT: THE MILITARY WAY

Lessons from the military can enable corporate firms to efficiently and effectively manage their employees.[8] The first determining factor is the "span of control." The term "span of control" means the number of managers required for supervision and the number employees a manager should supervise. This term is a *Civil Air Patrol* jargon. Span of control defines a manager's efficiency. If he/she has to supervise more people then he/she will not be able to supervise his/her subordinates effectively, consequently affecting the productivity of the employees. Also, if a manager has less people to supervise, this will lead to a laid-back attitude in the manager, reflecting the same attitude in the employees he/she is supervising. There is an apt figure for the number of managers required for supervision and for the number of employees a manager can supervise, depending on the industry type and responsibilities of the manager.

"Unity of command," as the term indicates, every employee should have only one boss. This principle is strictly followed in the military. Unity of command prevents overlapping of responsibilities, which reduces confusion among the managers and employees. Also, reporting and communicating with several supervisors can prove to be frustrating for the employees. This is because an employee finds it easy to understand and accommodate as per the commands from a single supervisor. More supervisors would require more adjustment and can create more confusion.

---

[8] Anonymous. (April 20, 2007). Management Principles Corporate Learning Course "Team Building" Block Seminar 3.4. Retrieved from http://www.capmembers.com/media/cms/Lesson_34_Management_Principlesdoc_85DA319EF79EF.pdf (accessed on July 31, 2014).

THE MILITARY PRINCIPLE OF ECONOMY OF FORCE

The principle of economy of force in the military, laid by Widmer, states that resources positioned should suitably correspond to the terrain and traits of the enemies. Describing the importance of the economy of force, Napoleon stated, "The first task of any general is to work out what has to be done. The next is to determine whether he has the resources to overcome any obstacles that the enemy can put in his way."[9]

Lessons that the corporate firms can derive from the aforementioned maxim is that firms should make sure that the goods/services they provide are in alliance with what the customers need or want. In many cases, it has been found that firms focus on producing high quality goods or cheap goods, and do not consider the quality aspect of the goods that the customers demand for. An example for this is the launch of Sony and IBMs' Personal Digital Assistants (PDAs). PDAs are basically mini versions of personal computers (PCs), such as notebook and palm laptops. Companies like Sony and IBM are known to have produced the most advanced technologies in the market. However, high quality and high-end features of their PDAs did not make them successful in the market. Consumers did not buy Sony's or IBM's PDAs, even though the technology used in them was ahead of their time. This was because the customers were not looking for such technologically advanced equipments. Sony and IBM, in this case, did not benefit from the *first mover's advantage*. In this competition, other companies like Apple entered the market and worked on the drawbacks that Sony's and IBM's PDAs had, and launched improved PDAs and captured different market segments by providing the customers with exactly what they wanted to buy.[10]

On the other hand, an example of provision of cheap quality goods is the goods manufactured in China. On the other hand, an example of provision of cheap quality goods is the goods manufactured in

---

[9] Widmer, H. (1980). Business lessons from military strategy. *McKinsey Quarterly, 2,* 65.

[10] Bayus, L. B., Jain, S., & Rao, G. A. (February 1997). Too little, too early: Introduction timing new product performance in the personal digital assistant industry. *Journal of Marketing Research, XXXIV,* 50–53.

China as perceived by Indians. In sample study conducted by the author, China is perceived to providing consumers with less-durable and low-quality products. Due to high price sensitivity of Indian consumers, Chinese phones initially flooded the Indian market with cheap replicas of high end mobile phones. Initially, customers bought these Chinese phones, but later realized that the phones had problems with parts and warranty. Seizing this opportunity, Micromax started offering low-cost smartphones in Indian market. Today, Micromax is number one cellphone vendor with a market share of 16.6 percent.[11] Today, Nokia is one of the most common names in rural as well as urban Indian mobile phone markets.

The aforementioned examples indicate lessons that firms should learn before introducing a new product in the market. It is essential that a firm thoroughly understands the markets and consumer needs, and the types of want it desires to cater to, and then, accordingly, manufacture goods.

Making Initiatives

A lesson learnt from the military history has now become the new mantra of management.

Sun Tzu stated, "Speed is the essence of warfare. Exploit the enemy's unreadiness, move by unexpected routes, and attack unguarded positions."[12]

This statement indicates that firms must act faster than their competitors. Firms must not wait too long to implement potentially successful ideas or plans. In case of introducing a new product variant, firms can be tempted to wait for some period, as they want to prevent it from competing with their flagship products. Firms like to wait till their previous stocks are over, and then launch a new product. However, firms tend to forget that their competitors may capitalize

---

[11] Khan, Danish. (2014). Micromax overtakes Samsung to become leading India handset vendor in Q2. *The Economic Times*. Retrieved from http://articles.economictimes.indiatimes.com/2014-08-04/news/52428502_1_micromax-smartphone-segment-phone-market (accessed on August 4, 2014).

[12] Tzu, S. (2009). *The art of war: The oldest military treatise in the world*. Part 6. (Lionel Giles, Trans.). New Delhi: Radha Press.

the same idea and capture the market. Studies indicate that ideas and innovation can enable creation of a new market, and pull demand for products.[13] Firms need to quickly assess the potential of their product in the market. If the market potential for the product is high then firms must act quickly in launching the product and become the pioneer.

Operations Planning

Operations planning is about assigning the resources and tasks for a mission. The planning process can be defined in the form of four phases in the Australian Defense Force Planning tenet.[14]

- Mission analysis
- Course-of-action development
- Course-of-action analysis
- Decision and execution

The first phase can be explained as identifying and defining the desired outcomes. Second phase describes the defined tasks to achieve the mission. Third phase can be explained as an identified set of actions, required to achieve the assigned tasks. Finally, the fourth phase indicates the decision making, followed by command for its execution by the commander-in-charge.

The same process can be followed by the corporate firms for their operational planning. The chief executing officer (CEO) needs to identify the organization's mission. Once the organizational mission is defined, a plan of action needs to be developed. This would include identifying the tasks involved in achieving the mission. This will be followed by assigning the tasks to the managers. Finally, the manager delegates the task to his/her subordinates, and gives them instructions on how they should execute their assigned duties.

[13] Anonymous. (February 2009). *The potential of market pull instruments for promoting innovation in environmental characteristics*. European Commission Directorate General Environment. Retrieved from http://ec.europa.eu/environment/enveco/innovation_technology/pdf/market_pull_report.pdf

[14] Aberdeen, D., Thiebaux, S., & Zhang, L. (2004). Decision-theoretic military operations planning. ICAPS-04 Proceedings, 402-411, Retrieved from http://www.aaai.org/Papers/ICAPS/2004/ICAPS04-047.pdf (accessed on July 25, 2014).

Professional Training

The military excels at providing the officers with the required training and certifications. Providing training for development of individual expertise is the fundamental factor that can enable officers to acquire the required specialized knowledge. Also, training is responsible for an individual's career and personal growth. Many firms do not provide their employees with advanced knowledge and expect them to learn the skills before joining them. On the other hand, if firms provide their capable and efficient employees with the required training, they will gain knowledge and try to give their best back to the firm. This will also lead to a low attrition rate, as individuals feel promised about their growth and progress in the firm. Also, training develops trust in the employees for the firm they work, and it brings a sense of responsibility in the employees and inculcates positive values, such as team work and integration. Timely provision of training is an important factor that leads to high employee retention rate.

Coordination

Coordination in the military is extremely effective. Great care is taken and huge investments are made to facilitate it. This is because one wrong message or no message can cause harm not only to the security of the troop, but also to the whole nation. The military uses various plans, signs, and control measures to co-ordinate with different units.

Coordination of this level and intensity will not be required for the firms. However, a lot can be adapted from the military. Firms should have all their units and security systems well connected. Efficient systems should be implemented for coordination and communication between various parties involved in the business.

For example, in case of general insurance, the insurance company should be well connected with the customer, claim investigators, third party insurance company, authorized garage, towing organizations, etc. Effective coordination within and outside the firms is an indicator of a firm's efficiency. Efficient services will attract more customers, leading the firm to be a strong player in the market.

Networking

The whole military system is well networked. Networking in a firm is an important tool for acquiring and sharing knowledge. Efficient networking systems enable efficient sharing of knowledge within the members of the firm. It is important that knowledge and information is passed on to everyone in the firm. Knowledge and efficient communication can prove to be problem-solving resources. Networking leads to a channel of information push and information pull in an organization.

Good Information Technology (IT) Infrastructure

Good IT infrastructure in a firm indicates efficient access to information across all vertical and horizontal integrations. With the help of IT information on inventory, distribution, and supply, statuses can be tracked efficiently. Within the firm, it becomes easy for people to find and identify other members of the firm. Not only this, but also, with the help of efficient IT systems implementation, reaching out to the customers becomes easy and vice versa.

Don't Just Manage ... Lead

Leadership and management are two important concepts used in the military and in the business world. In general terms, management is associated with bringing together all the resources that are required to achieve the mission of the organization they are working for. On the other hand, leadership is getting people to work as per the leader's norms. A good leader can assure that the behavior of the employees is well guided. Leadership is driven by the achievement of ascertained objectives of the firm. Leadership is about getting people into action, guiding them, and motivating them to accomplish the organization's desired outcomes.

A leader in the military is termed as a CEO in the business world. Tzu described the importance and responsibility of a leader as: "Now the general is the pillar of the State: if the pillar has mastered all points of war, the State will be strong; if the pillar is defective, the State will be weak."[15]

[15] Tzu, S. (2009). *The art of war: The oldest military treatise in the world.* Part 6. (Lionel Giles, Trans.). New Delhi: Radha Press.

Ancient Thai History and Battlefield Strategies

The battlefield strategies of ancient Thai warriors were believed to have evolved in a unique traditional Thai way, and were not derived from other parts of the world. The ancient Thai battlefield strategies considered certain constructs:

- **Planning with Objectives**
  The Thai Army always followed the approach of planning with objective. Planning is one of the most significant processes, which an army uses to determine its relationship with the environment and its strength against its enemy. It is a synthesis of collecting information, making decisions, helping management formulate objectives, and choosing the pattern of action to achieve these objectives.
- **Administration and Control**
  In order to ensure that the planned course-of-action is followed, administration and control need to be handled carefully. The administration of an army is generally concerned with organizing and supervising its functions, and the control is concerned with the observation of results to see if battle plans are being carried out correctly.
- **Offensive and Defensive**
  The offensive warfare principle can be used by the ruler or general to defeat the enemy. This approach initiates the operations or the attack on the enemy and shapes its situation. On the other hand, the defensive strategy should be considered as a proactive exercise, where one can build strength for overcoming their weaknesses. The defensive strategy mainly requires considerable self-discipline for implementation purposes.
- **Concentration or Mass**
  The strategic principle of concentration or mass, basically, refers to the mobility, collaboration, and reunion of scattered forces so that it can substantially form strong troops to effectively attack the enemy. It is basically related to the organization of the army by mobilizing and concentrating other supporting and reserve forces to a specific strategic location, aimed at defeating and destroying the enemy for a complete victory.

- **Maneuvering**
  Maneuvering is concerned with the clever and skillful changing of a situation so that one benefits from it. The objective of maneuvering is to change the place of the army to advantageous areas, where one can easily defeat and destroy the enemy, using the least resources. The fundamentals of this principle are flexibility and fluidity in organizing the forces, readiness and quickness in supporting the forces, and effectiveness in commanding the forces.
- **Surprise**
  Surprise here mainly refers to the defender's unreadiness, caused by one or more mistaken estimates about the enemy, that is, when, where, or how the enemy would strike. The surprise attack generally reduces the enemy's morale, especially when a surprise attack is successful and disruptive.
- **Cautiousness**
  Cautiousness is another important aspect as it brings along a protected, secured, and safe environment for the army or country. Being cautious means being detailed and thorough in planning and execution, relative to the enemy.[16]

Clausewitz proposes that the best way to overthrow an enemy is to disarm him/her. As long as a business exists, competition from others will continue to exist. Global businesses need formulation of new growth strategies in order to survive the escalating issues of competition. Growth in business refers to increasing the size and the capability of the firm with time. Competitive strategies are also useful in determining how others are strategizing themselves.[17] Just as in war, businesses also need to disarm their enemies. To effectively do this, they require motives and valuable activities that prevent them from being compromised. The strategy of disarming an enemy

---

[16] Low, Sui P., & Chuvessiriporn, C. (1997). Ancient Thai battlefield strategic principles: Lessons for leadership qualities in construction project management. *International Journal of Project Management, 15*(3), 133–140.

[17] Hitt, M. A., Ireland, D., & Hoskisson, R. E. (2009). *Strategic management: Competitiveness and globalization: Concepts and cases.* Ohio, USA: Cengage Learning.

or competitor is to gain information about him. Competitors can be overpowered by developing new and unique strategies that are beyond their reach. Internalization is an important aspect of strategy formulation, which is made up of important elements, such as market drivers, cost, competitive and government drivers.[18]

TACTICS AND STRATEGY: ELEMENT OF SURPRISE

We go back to the time of the year when India awoke to freedom and had to fight its first war with Pakistan over Jammu and Kashmir. This was the first Kargil War, fought in the year 1948 and is considered to be the most remarkable time in the history of Indian Army. This war throws light on the life of Great General Thimayya.

Gen Thimayya's tactical maneuvers and strategies during the war helped our forces to save Kargil. It was under his leadership and command that the Stuart Light tanks were deployed to Zojila Pass. Zojila Pass commonly known as "Path of Blizzards" is approximately 11,000 feet above sea level. It had no vegetation to conceal troop movements, therefore the Pakistani positions on the steep mountain slopes were impossible to hit either from the air or by land. At the same time, there was a shortage of Indian troops to carry out a full-fledged attack.[19]

In February 1948, the Pakistani military began operation with the objective to seize Skardu, Kargil, Zojila pass, and Leh in Ladakh. The Pakistani forces surrounded the Kargil, Dras and Skardu. Their ambush destroyed two Indian Battalions. Subsequently, Pakistani army seized Kargil, Dras, and Skardu. It was the annexure of Leh that opened up trails for the Pakistani raiders.

(Box continued)

[18] Karami, A. (2007). *Strategy formulation in entrepreneurial firms*. Aldershot, Hampshire: Ashgate Publishing.
[19] Singh, Lt Gen B. Battles for Zoji-La and Namka Chhu. http://www.indiandefencereview.com. [Online] May 5, 2014. http://www.indiandefencereview.com/spotlights/battles-for-zoji-la-and-namka-chhu/

(Box continued)

> The Pakistani forces caught the Indian arms off-guard. However these tactics did not surprise Major General Thimayya. It was as if he was already prepared for such an adverse situation. He immediately came up with a truly ingenious and bold plan, and issued instructions to move tanks from Akhnoor in the plains to Zojila through various passes. General Thimayya led the attack from the front and succeeded in capturing Dras, Kargil and Leh. He led the brigade by boarding the forward most tank and was supported by a brigade of Stuart Light Tanks of the 7th Light Cavalry. His leadership and ability to stay calm in moments of extreme stress helped his team to drive out the raiders.[20]
>
> The combos moved only during night and after their arrival in Srinagar, the city was placed under curfew.[21] Pakistani intelligence was unaware of these movements and the preparations. Pakistan was amazed by the arrival of the tank. After hunting and killing many Pakistanis from their caves, the tanks moved to Kargil at Matayan which resulted in capturing Kargil and clearing the linkup with Leh.[22]
>
> Ultimately on November 24, the most remarkable, rather, the first unofficial war came to an end with the success of capturing Kargil and making Ladakh safe. Gen Thimayya's indelible courage in the moments of extreme difficulties and his ability to deliver results through the element of surprise allowed India to win one of the most historic wars. He has always been respected and loved by all ranks and was undoubtedly a born leader. His decision of moving the tanks to Zojila through different passes

(Box continued)

[20] Sarkar, B. (1999). *Kargil war: past, present, and future*. s.l.: Lancer Publishers.

[21] Biswas, K. The general and his tanks! http://defenceforumindia.com. [Online] Retrieved on June 26, 2011 from http://defenceforumindia.com/forum/indian-army/23011-general-his-tanks.html

[22] Guruswamy, M. The Saga of Zojila. http://expressindia.indianexpress.com. [Online] Retrieved from http://expressindia.indianexpress.com/ie/daily/19981023/29650324.html (accessed on October 23, 1998).

(Box continued)

> was itself a Herculean task that emphasizes his indomitable leadership qualities. Even though the Pakistani raiders had a surprise element to their charge, he was still able to stay focused and came with an out of the box innovative solution. India's victory is owed to the bravery of men on the borders such as Gen Thimayya, who showcase the ability to operate in tough situations and yet deliver favorable results.
>
> Army continues to emphasize on using element of surprise as a tactic and a strategy to beat the adversary. However, one cannot afford to be surprised, so, preparedness for the surprise is another success factor for the army.

7

WORK–LIFE BALANCE

INTRODUCTION

Maintaining a balance between work and family life is at an individual's residual. One has to juggle between the two to maintain a balance, and set the priorities right to be able to do so. Although work–life balance is perceived as an individualistic task, its repercussions are not. Maintaining work–life balance can be stressful, as individuals try to maintain and manage the most important aspects of their lives. In a holistic view, work and home are interconnected. Thus, if one is not happy with one's work, its effects are seen in one's performance at home. Similarly, if one is not happy with the environment at home, its effect will be seen in one's performance at work. Therefore, it is no longer safe for an organization to think that its employees' work–life balance is their individual problem. The impact of work–life conflict extends to an individual's spouse, children, supervisors, and subordinates. It has become imperative that organizations provide their employees with facilities that are conducive to work–life balance.

Work and family life are two mutually exclusive parts of a person's life. However, it is nearly impossible to separate one from the other. They interfere with each other at many instances, and the happiness and stability of one sphere of life directly affects the efficiency in the other. A man living a balanced family life will find it easier to cope with the ever mounting pressure at the workplace; however, a man with a problematic family life would have a disturbed state of mind and would not be able to perform at the desired levels of efficiency. On the other hand, a man with a tiring job would find it difficult to spend quality time with his family. This would, again, in turn, affect the efficiency. It is difficult to keep the two mutually exclusive events apart from each other; however, if both are kept balanced then productivity increases. A person with a happy and excited mind is more attached to both his/her work and family. Hence, he/she gives the best to be as useful as possible to the organization.

Organizations are well aware of the importance of work and life balance. An unhappy employee has a negative influence on the others, the efficacy decreases, and there is a possibility of the employee shifting to another workplace because of the unsatisfactory working hours. The organizations are well aware of this, and, hence, try to provide sufficient family time to the employees. The armies all over the world recognize the bond that a soldier shares with his family, and, hence, take care of both of his/her professional and personal needs.

WORK–LIFE INTEGRATION

Kofodimos defined work–life integration as

> [A] satisfying, healthy and productive life that includes work, play, and love; that integrates a range of life activities with attention to self and to personal and spiritual development; and that expresses a person's unique wishes, interests and values. It contrasts with the imbalance of a life dominated by work, focused on satisfying external requirements at the expense of inner development, and in conflict with a person's true desires.[1]

Huffman, Culbertson, and Castro stated that the perception of the members of the military about his/her family-friendly environment is

---

[1] Kofodimos, R. J. (1993). *Balancing act: How managers can integrate successful careers and fulfilling personal lives* (1st ed.). Ann Arbor, Michigan: Wiley, Jones and Sons, Incorporated.

directly related to his/her physical fitness, efficacy beliefs, and intentions to stay in the military. Not only that, it has been proven that family-friendly environments help lessen the negative effects of the work–life conflict.[2]

PROBLEMS CREATED BECAUSE OF LACK OF WORK–LIFE BALANCE

Work–life conflict has become a part of every employee's life. Work–life conflict arises when an individual is not able to maintain a balance between work and home, because of spillovers of needs from both the fronts. Studies have indicated that work–life conflict has a direct link with many negative consequences on an individual. Frone and Russell indicated in their work that work–life conflict leads to lower well-being in individuals.[3] It also leads to lower job satisfaction among employees.[4] Allen, Herst, Bruck, and Sutton indicated in their work that work–life conflict also led to poor health and increase in alcohol intake.[5] As a result of these negative effects on the employees, the organization, as a whole, faces negative consequences of the work–life conflict. Jex has explained from his study that work–life conflict increases turnovers and decreases work performances by employees.[6]

In case of members of the military, studies indicated that work life was directly related with the physical fitness, efficacy beliefs, and retention rate of the officers. Conflict in work life has a negative

---

[2] Huffman, A. H., Culbertson, S. S., & Castro C. (2008). Family-friendly environments and US army soldier performance and work outcomes. *Military psychology. 20*(4), 253.

[3] Frone, M. R., Russell, M., & Barnes, G. M. (1996). Work–family conflict, gender, and health-related outcomes: A study of employed parents in two community samples. *Journal of Occupational Health Psychology, 1*, 57–69.

[4] Adams, G. A., King, L. A., & King, D. W. (1996). Relationships of job and family involvement, family social support, and work–family conflict with job and life satisfaction. *Journal of Applied Psychology, 81*, 411–420.

[5] Allen, T. D., Herst, D. E. L., Bruck, C. S., & Sutton, M. (2000). Consequences associated with work-to-family conflict: A review and agenda for future research. *Journal of Occupational Health Psychology, 5*(2), 278.

[6] Jex, S. M. (1998). *Stress and job performance: Theory, research, and implications for managerial practice.* Thousand Oaks, CA: SAGE.

impact on officers' physical fitness, efficacy, and retention in the organization.

The US military has been a good example in helping its members with solutions for maintaining a balance in their lives. It has implemented family-friendly policies to help its members have a balanced work and family life. Militaries, in general, provide on-site education, day care centers, youth and recreational services, and family accommodative leaves.

Similar policies are also implemented by corporate organizations. However, a major difference is that their employees' perception for the family-friendly policies would be different. In a corporate organization, most of these policies are dictated or influenced by the immediate supervisor. Therefore, a major amount of effectiveness of work–life balance policies depends on how the supervisors implement them.

- **Individual Self-Efficacy**
  Having a family-friendly environment is very important for the members of the military. Bhanthumnavin, in his study, has proved that there is a positive relationship between individuals' efficacy beliefs and family-friendly work environments.[7] As the military provides its members with a balanced work and personal life, they are able to live through the hardships without seeing their families for long periods.
- **Employee Turnover**
  Employee turnover ratio has been a matter of concern for many organizations. This is because employee turnover comes at a cost. It is pervasive and occurs at a high cost to the organization. Employees, who have newly joined an organization, undergo various training programs. It takes time for an employee to get accustomed and trained as per an organization's requirements, and the company bears the cost to provide the training and for paying the salaries. If the employee turnover rate in an organization is very high, it

---

[7] Bhanthumnavin, D. (2003). Perceived social support from supervisor and group members' psychological and situational characteristics as predictors of subordinate performance in Thai work units. *Human Resource Development Quarterly*, 14(1), 79–97.

loses out on the employees it trained, who with time would have become more productive. To find their replacement, organizations will have to bear the cost of recruitment and the cost involved in training new employees.

- **Family-Friendly Environment and Work–Life Interaction**
  Various studies have indicated the existence of a positive relationship between family-friendly environments and positive attitudes related to work.[8] Using the social exchange theory, researchers have explained that employees in a firm reciprocate in a way they think their organization is treating them. Therefore, if employees feel that their organization is supporting them, they will feel indebted to the organization and will try repaying it by improving their performance, and by staying with the firm for a long duration, in spite of conflicts at home.

A study by Bourg and Segal contemplated that family-friendly policies in the military buffers the relationship between work–life conflicts and organizational outcomes.[9] Organizational policies providing support to the employees generate the willingness among them to perform their best (Figure 7.1). A supportive policy also moderates the stress and its impact on the employees' health, making them perform at par with their abilities.[10]

FACTORS CAUSING WORK–LIFE CONFLICT

- **Limited Resources**
  Every individual has a limited set of resources, in terms of time, money, psychological and physical energy. Individuals try to

[8] Huffman, A. H., Culbertson, S. S., & Castro C. (2008). Family-friendly environments and US army soldier performance and work outcomes. *Military Psychology, 20*(4), 253.

[9] Bourg, C., & Segal, M. W. (1999). The impact of family supportive policies and practices on organizational commitment to the army. *Armed Forces & Society, 25*, 633–652.

[10] Viswesvaran, C., Sanchez, J. I., & Fisher, J. (1999). The role of social support in the process of work stress: A meta-analysis. *Journal of Vocational Behavior, 54*, 314–334.

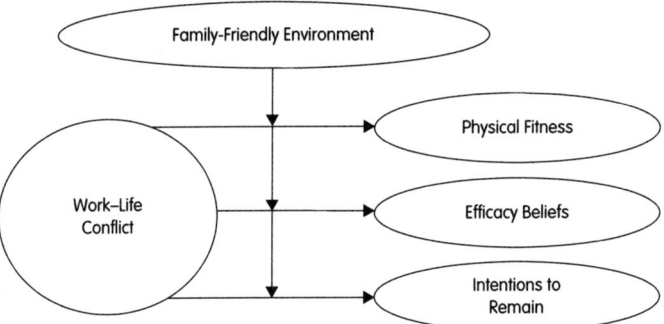

Figure 7.1 Factors Influenced by Family-Friendly Environment
Source: Huffman, H. A., Culbertson, S. S., & Castro, C. A. (2008). Family-friendly environments and US Army soldier performance and work outcomes, *Military Psychology, 20*(4), 253.

perform their multiple duties and roles using these resources. Individuals are not stressed when the resources meet their needs. However, when the available resources exceed the needs of an individual, stress begins. This mismatch leads to disturbances at the work and family fronts. The role strain hypothesis, propounded by Greenhaus and Beutell, suggests that strain and the associated stress has its spillovers in all the domains of an individual's life.[11]

- **Work and Personal Life Demands**
  There are two types of interferences that lead to work–life conflict in an individual's life. One is external and the other is internal. Internal interference would refer to factors such as tiredness and stress because of work. The external interferences would refer to factors such as long working hours.[12] Both these interferences create conflict in an individual's family life. In such a situation, individuals are not able to pay attention to work, and even family.

[11] Greenhaus, H. J., & Beutell, J. N. (January 1985). Sources of conflict between work and family roles. *The Academy of Management Review, 10*(1), 76–88.

[12] Carlson, D. S., & Frone, M. R. (2003). Relation of behavioral and psychological involvement to a new four-factor conceptualization of work–family interference, *Journal of Business and Psychology, 17,* 515–535.

- **Dependents**
  More number of dependents indicates more demands from the family end. Spending more time on family matters reduces the time one spends in working for the organization one is employed in, leading to lower work productivity. Nordenmark also found that the number of children in a family was directly related to the willingness and desire to spend more time at home.[13] This means that if more number of children an individual has, it reduces the time he/she wants to spend at work. However, because of work commitments, individuals cannot spend the required and desired amount of time on household matters leading to conflicts in one's mind, which is then reflected in their other facets of life.

- **Technology**
  Technology has made the world a small place. Anyone across the world can get in touch with you using the various available means of communication. Today, phones, e-mails, etc. have become the basis of our daily conversation. Thus, we are always glued to our gadgets. One can now work from home and keep in touch with everyone in just a click using the ETC (electronics and telecommunication) equipments. People are connected to everyone else so much that they do not really find time for themselves. Technology has proven to be a boon on one hand, as it imparts flexibility to the employees; but on the other hand, it steals away an individual's private space and peace. Constant calls and e-mails do not allow individuals to focus on something without being interrupted. Technology has changed the social norms. Today, if you do not attend a call or respond to mail, it is considered offensive. To prevent this, people stay glued to their phones and personal computers (PCs), leading to excessive usage. Studies have indicated that excessive usage of mobile phones leads to psychological

---

[13] Nordemark, M. (2002). Multiple social roles—a resource or a burden: Is it possible for men and women to combine paid work with family life in a satisfactory way? *Gender, Work and Organization, 9,* 125–145.

disorders.[14] Thus, technology, today, plays a major role in contributing to the work–life conflict.

CONSEQUENCES OF WORK–LIFE CONFLICT AT WORK

Work–life conflict of employees has an impact on the organizations too. These impacts are primarily seen in the form of absenteeism, lack of job satisfaction, psychological imbalances, and health effects on the employees.

- **Withdrawal Behaviors (Absenteeism)**
  Thomas and Ganster stated that work–life conflict had a positive relationship with work absenteeism and other withdrawal behaviors.[15] Studies have also indicated a direct relationship between employees' turnover intentions. It was also observed that people who had low levels of career involvement demonstrated more behaviors of withdrawals.[16]
- **Lack of Satisfaction**
  Work–life conflict lowers an individual's satisfaction in terms of personal and professional life.[17] Work–family interference also affects marital satisfaction. Work–life conflicts affect satisfaction in life, family, and work. It leads to unrest in the family, and reduced productivity and motivation on the work front.
- **Psychological Effects**
  Conflict in work and personal life leads to psychological distress.[18] Many cases have been reported where individual

---

[14] Anonymous. (February 26, 2012). Phone obsession can cause psychological problems. Retrieved from http://articles.timesofindia.indiatimes.com/2012-02-26/computing/31101278_1_nomophobia-mobile-phone-lookout-mobile-security

[15] Thomas, L. T., & Ganster, D. C. (1995). Impact of family-supportive work variables on work–family conflict and strain: A control perspective. *Journal of Applied Psychology, 80*(1), 6.

[16] Jones, F., Burke, J. R., & Westman, M. (2006). *Work–life balance—A psychological perspective*. Hove, Sussex: Psychology Press, Taylor and Francis.

[17] Carlson, D. S., & Kacmar, K. M. (2000). Work–family conflict in the organization: Do life role values make a difference? *Journal of Management, 26*, 1031–1054.

[18] Major, V. S., Klein, K. J., & Ehrhart, M. G. (2002). Work time work interference with family, and psychological distress. *Journal of Applied Psychology, 87*, 427–436.

involved in work–life related conflicts often face depression. This has been a common feature in men and women both.
- **Health**
Recent studies have indicated an interrelation between work–family conflict and negative physical health effects.[19] Lee found that spillovers from the demands from home and work led to the following symptoms of stress: headaches, weight loss and weight gain, giddiness, insomnia, etc.[20] In a few cases, strain caused by work–family interferences also led to heart diseases, loss in appetite, low energy levels, anxiety, nervousness, tension, etc.[21]
- **Other Consequences**
Excessive intake of alcohol was another major negative consequence of work–family conflict. Other consequences were low organizational commitment level and low work productivity. Besides, on the home front, work–family conflict also led to destructive parenting and child behaviors. It has also affected individual cognitive ability to pay attention and concentration in their tasks.

DIFFERENT FORMS OF WORK–LIFE CONFLICT

Work–life conflicts exist because of competing demands, which stretch a person from both ends. There are several reasons for this, such as overworking, increased work pressure, etc. However, these conflicts exist in certain forms.

- **Time-Based Conflicts**
Time-based conflict arises when the time invested in work interferes with the time required to fulfill the necessities of

[19] Frone, M. R., Russell, M., & Cooper, M. (1992). Prevalence of work–family conflict: Are work and family boundaries asymmetrically permeable? *Journal of Organizational Behavior*, 13(7), 723–729.

[20] Lee, J. A. (1997). Balancing elder care responsibilities and work: Two empirical studies. *Journal of Occupational Health Psychology*, 2, 220–228.

[21] Allen, T. D., Herst, D. E. L., Bruck, C. S., & Sutton, M. (2000). Consequences associated with work-to-family conflict: A review and agenda for future research. *Journal of Occupational Health Psychology*, 5, 278–308.

the family. This conflict can be attributed to the simple zero sum nature, which states that an individual cannot be present at both places at the same time. So, if an individual is working late, he/she will have to end up missing the quality time he/she could have spent with the family. Another form of this conflict is the preoccupancy with work, when spending time at home. This is equivalent to not being at home, as majority of the attention is at work.

- **Strain-Based Conflicts**
  Strain-based conflict implies strain in the family domain because of spillovers from the work front. If an individual had a bad, or a busy day at work then in this case the frustration or the tiredness will show up in his/her activities when at home. For example, if an individual gets so tired that he ends up sleeping early and reaching work early, leaves him isolated from the family interactions and activities like playing with the kid, helping his wife, etc.
- **Behavior-Based Conflicts**
  Behavior-based conflicts indicate the spillovers from behavior at workplace that is incompatible at home. This happens when an individual spends a lot of time behaving in a particular way and gets conditioned to it. However, that behavior of the individual disrupts the family environment. For example, if an individual is a supervisor, who is used to getting work done by ordering and commanding others, practices the same behavior when at home, it can hurt the feeling of people around him/her.

Work–Life Conflict at Home

Work–life conflict has a major impact on the personal life of an individual. Personal life can be divided into two parts; first, being the family, and the second, being oneself. Work–life conflict intrudes in the family space of individuals, preventing them from spending quality time with their loved ones, including friends. Besides the loved ones, there is also another important aspect of an individual's life, that is, time for oneself. It is essential for individuals to spend some time with themselves, doing things the way they like.

Spending time with oneself is as important as spending time with family, as it helps a person in gaining peace and to unwind. Work–life conflict can jeopardize various domains of an individual's life.

Effect of Work–Life Conflict on Families

The negative effects of work–life conflict are not only seen on the individuals, but its radiations also affect the entire family. Work–life conflict in an individual's life can also affect his overall family's performance. If the father in a family does not give time to his family, a lot of things go unattended. This leads to further problems between the spouses, creating more tension. This could be in the form of less attention to the spouse and children, leading to unfulfilled emotional needs of people. This baggage is then carried on to work by the employee, thus affecting his work performance and efficacy levels.

Work–life related conflicts lead to low marital satisfaction, leading to increased conflicts between the spouses. This further leads to instances of family strife and divorces. Also, conflicts faced by one of the partners have an effect on the other partner.[22] Besides this, there are some other impacts of work–life conflict. With more and more couples getting involved in dual careers, they end up spending less time with each other and their children. Many fathers do not get to see their children for weeks and months. Work related conflicts also lead to parental problems, creating further unrest among all the family members. Instances of juvenile delinquency and violence have also been observed.[23] All of these lead to increased stress levels and tensions, and have an impact on the day-to-day operations of the family. Parents play the most important role in the upbringing of a child. If this is not taken care of, children often get depressed, or adapt to unhealthy habits and environments. This can be seen in the form of staying in the company of wrong people, not studying, etc.

[22] Quick, J. D., Henley, B. A., & Quick, C. J. (2004). The balancing act—At work and at home. *Organizational Dynamics, 33*(4), 426–438.

[23] Allen, T. D., Herst, D. E. L., Bruck, C. S., & Sutton, M. (2000). Consequences associated with work-to-family conflict: A review and agenda for future research. *Journal of Occupational Health Psychology, 5*(2), 278–303.

UNDERSTANDING THE ROLE OF FAMILY

It is an idealistic saying that one should not mix one's private and personal life. However, how much can this hold true? Both these aspects of an individual's life are interconnected. One does affect the other. An individual leading a happy family life is more productive at work than a person whose family life is not at ease.[24] Family and friends are a source of emotional empathy, understanding, motivation, guidance, and information. Therefore, if the work–life of an individual is taking a toll on his/her personal life, this will affect his/her performance at work. Spouses play an important role in an individual's life. As per the conservation of resources model, propounded by Hobfall, a spouse is considered as an additional resource, who provides support in stressful situations.[25] Grandey and Cropanzano also suggested that having a partner reduced the self-supported stress of an individual.[26]

More Women Entering the Corporate Arena

Like military, other organizations have also opened their gates for women's participation. Initially, employment of women was only limited to administrative and clinical profiles. They were strictly not allowed to participate as members of the combat team. However, this scenario has now changed. Women, today, are treated equally in the military. Common entrance examinations and tests are conducted for both the genders. Women are allowed to be a part of the combat team.

The aforementioned scenario is also applicable to the corporate world. Initially, women were not given the top positions in organizations. Later, as policies changed, so did the role of the women in the society; women started to work and were considered competent. Today, we can find many top organizations headed by women.

[24] Ibid.
[25] Hobfall, S. E. (March 1989). Conservation of resources: A new attempt at conceptualizing stress. *American Psychologist, 44*(3), 513–524.
[26] Viswesvaran, C., Sanchez, J. I., & Fisher, J. (1999). The role of social support in the process of work stress: A meta-analysis. *Journal of Vocational Behavior, 54*(2), 314–334.

However, the ratio is still low. Not many women are seen at the top positions in corporate organizations. Various studies have indicated that women are still marginalized and prevented from promotion and remuneration. There are evidences of gender-based discrimination in corporate organizations across the globe. A study by Grandey and Cropanzano explained that women tend to experience higher levels of stress and work–life conflict spillovers than men.[27]

As cited in the International Labor Organization (ILO) 2004 Report, "women absorbed less than 15 percent of chair as corporate board members in the USA, UK, Canada, Australia, and many European countries." This ratio in some Asian countries is as low as 0.2 percent.[28] Bilimoria explained that more women as the board members represented a potential increase in the career opportunities for women.[29] Another attribute to this, was that women chose not to opt for challenging positions because of their family commitments. With the increasing financial pressures on individuals, more and more women have and are still joining the economic work force. Women work hard but still do not get the stature they deserve. This is where an organization can play the role of a buffer. There is a need for developing separate policies for women, as they play a major role in households.

OTHER FACTORS THAT INCREASE THE WORK–LIFE CONFLICT

There are other indirect factors that are created because of work–life conflict and they further increase the imbalance.

- **Consumption of Alcohol and Other Products**
  Alcohol has been termed as one of the most harmful causes of work–life conflict, disturbing the family life. It has been found

[27] Grandey, A. A., & Cropanzano, R. (1999). The conservation of resources model applied to work–family conflict and strain. *Journal of Vocational Behavior, 54,* 350–370. Retrieved from http://www.idealibrary.com/ (accessed on July 31, 2014).

[28] Elder, S., & Schmidt, D. (2004). *Global employment trends for women, 2004.* Employment Analysis Unit, Employment Strategy Department, International Labour Office. Geneva, Switzerland.

[29] Bilimoria, D. (2006). The relationship between women corporate directors and women corporate officers. *Journal of Managerial Issues, 18*(1), 47–62.

that people tend to consume more amount of alcohol when stressed with work.[30] Consequently, excessive amount of alcohol impairs one's senses to act rationally. The individual can start behaving in an uncontrolled manner. This not only hampers his/her work, but it also hampers the state of mind of the family members, and, also, the productivity of the organization.

- **Unsound Sleep**
  Stress has been known as the major cause for disturbances in sleep, such as unsound sleep and insomnia. Short-term sleep disturbances may not cause problems, but long-term sleep disorders can be a matter of concern. Consistent lack of sleep can cause problems in concentration, dealing with tasks at work, and, also, decision making. This can hamper an individual's career and peace at home. The spillovers from one domain eventually pass on to the other.

- **Increased Traveling**
  With increasing globalization, traveling of people from place to the other has substantially increased. One problem with this could be that the individual feels distant because of his consistent physical distance with the family. This reduces his/her dependency on the family and vice versa. Frequent traveling of a spouse can lead to interruptions in the regular emotional reassurances and the social support provided by the family. The second problem can be in the form of increased work pressure on one of the spouses because of lack of presence of the other. There are so many routine activities that a spouse, in general, can contribute to. In case of absence of one of the spouses, the other spouse can be burdened with extra or unmanageable responsibilities. This can lead to frustration and fights between the couple, leading to disturbances in the family integrity.

- **Organizational Cultures**
  Corporate organizations are known for their unhealthy work cultures. Unhealthy work culture includes working for long

---

[30] Allen, T. D., Herst, D. E. L., Bruck, C. S., & Sutton, M. (2000). Consequences associated with work-to-family conflict: A review and agenda for future research. *Journal of Occupational Health Psychology, 5*(2), 278–303.

hours, lack of flexibility to work from home, workplace politics, and fierce competition among colleagues. Such work environments are common, but are detrimental to the well-being of the employee's family life. No matter how hard one tries, one will feel the discomfort of the unhealthy work culture, if he/she is a part of one. Such work environments do not allow the employees to come out of the work mode, hampering their performance at work. They do not provide the employees with the time to relax, unwind, and to get prepared for work again.

- **Increased Accessibility**
Individuals with access to the Internet via portable gadgets anytime, such as laptops, notepads, and mobile phones, have become available to and accountable to the whole world. On one end, it is the best discovery made to ease the operations and communication in organizations. However, on the other end, e-mails and mobile phones have ruined the peace and personal space of an individual. Internet, today, can enable employees to work from wherever they are. Employees are accountable to check their e-mail boxes often and to reply to the concerned mails. Such habits intrude in the family space of an individual and take him/her back to work mentally in a few seconds. Although e-mails are better than calls and texts on cell phones, an individual has the option of checking and replying to the mails at his/her discretion.

DEALING WITH WORK–FAMILY CONFLICT: BALANCING STRATEGIES

Experts have come up with strategies that can help employees maintain a work–life balance. The first step toward this is to develop the understanding that maintaining the balance majorly depends on an individual.

- **Selection, Optimization, and Compensation (SOC)**
The SOC model was propounded by Baltes and Heydens-Gahir. They suggest that the SOC model can prove to be beneficial to those who are facing an imbalance in their work

and life.[31] This theory suggests that in adulthood, individuals should construct goal systems on the basis of optimization and compensation. As per the result of the study, individuals who followed the SOC model experience less work–life conflicts. They fared higher scores in their success in professional life and life in general.[32]

- **Playing an Active Role**
Ellen Kossek and Raymond Noe suggested that individuals should play an active role in managing their family and work roles and responsibilities. This approach has been explained as boundary management. Individuals should set boundaries for themselves by distinguishing between the role and responsibilities assigned, and the expectations from work and home front. This way, individuals will be less burdened and will execute their duties based on the priorities.[33]

As an addition to this, *Christena* demonstrated two strategies to manage work–life balance. In the first strategy, individuals should lay clear boundaries in work and life, in general, with no overlaps. In this strategy, individuals should try to maintain two different boundaries between work and general life by keeping them distinct. This separation will help individuals in mixing two different domains with each other, thus reducing the related stresses. The second strategy is that individuals should follow no boundary to manage between both the domains. In this strategy, individuals need not define the boundaries for themselves, but instead, they can make themselves available to either of the boundaries as required.[34]

---

[31] Baltes, A. B., & Heydens-Gahir, A. H. (2003). Reduction of work–family conflict through the use of selection, optimization, and compensation behaviors. *Journal of Applied Psychology, 88*(6), 1005–1018.

[32] Wiese, S. B., Freund, M. A., & Baltes B. P. (2000). Selection, optimization, and compensation: An action related approach to work and partnership. *Journal of Vocational Behavior, 57*, 273–300. Retrieved from http://dtserv1.compsy.uni-jena.de/__C1256E7F00217282.nsf/0/D872677C7504CB34C1256EA1002A08E0/$FILE/Wiese.pdf

[33] Kossek, E. (1999). Work–family role synthesis: Individual and organizational determinants. *International Journal of Conflict Management, 10*(2), 102–129.

[34] Nippert-Eng, C. (1996). *Home and work: Negotiating boundaries through everyday life.* Chicago: University of Chicago Press.

- **Flexible Work Environments**
  Flexible work environments bring feelings of motivation and dedication to work. It also enables the employees to utilize their time in the most optimal way, leading to a balanced work–life and reducing conflicts.[35] Flexible work culture provides major support to the parents, who have more demands to meet on the home front. However, there can be downsides to this as well. Studies have also indicated that flexible environments can also lead to more stress, as this often extinguishes the boundaries between the work and family domain. This can lead into more instances of work–life conflict.[36] It is, thus, important that organizations provide add-on flexibility to their employees on the basis of an individual's necessity.

- **Organizational Support**
  Organizations can provide support to their employees in the form of support from their immediate supervisor. Emotional support at work helps individuals balance work, as it becomes a source of increasing energy and motivation at work. A supportive supervisor can boost employees' energy levels, provide emotional support, and reinforce employees' positive image.[37] Employees who are single and away from home could also receive counseling and guidance from their supervisors.

  This does not mean that the supervisor facilitates the work–life balance for the employees, but it refers to helping out the employees in accomplishing their tasks at workplace. This will give them sufficient time and peace of mind to attend to the domestic needs. This also helps employees in not carrying their work-related issues to home, as they are aware that they will receive help from their supervisors.

---

[35] Hill, E. J., Ferris, M., & Martinson, V. (2003). Does it matter where you work? A comparison of how three work venues influence aspects of work and personal/family life. *Journal of Vocational Behavior, 62*, 220–241.

[36] Desrochers, S., Hilton, J., & Larwood, L. (2005). Preliminary validation of the work–family integration-blurring scale. *Journal of Family Issues, 26*(4), 442–466.

[37] Halbesleben, J. R. B. (2006). Sources of social support and burnout: A meta-analytic test of the conservation of resources model. *Journal of Applied Psychology, 91*(5), 1134–1145.

- **Family-Responsive and Friendly Organizational Culture**
  Besides providing emotional support and flexible work environment, organizations should also provide their employees with family-friendly organization culture. Organizational culture, which combines work and family role, should be adapted by the organizations, as this facilitates work–life balance for the employees. A family-friendly organizational culture provides a supportive environment, which provides emotional comfort in the form of understanding, advice, and recognition.[38] Family-friendly work culture creates a positive feeling toward organizations. This drives the employees to work better, improve their performance, and repay the organization in some way. This creates a win-win situation for both, the employers and the employees.

- **Self-Management**
  Self-management is about emotional competence. A study pointed out that people tend to experience less work–life related stress as they grow older. Older people, because of their experience, are able to manage work–life conflicts better. The younger generation needs to define their roles in their work and professional domain. They often tend to pay more attention to the role depending on their psychological involvement in them. At times, one goes to the extreme that one becomes prone to chronic fatigue, short tempered, disorientation, insomnia, etc.[39] This extreme behavior can disrupt family and work–life harmony. Individuals must try to maintain a balance between both the domains actively. Individuals need to realize that they will have to make constant active efforts to maintain a balance between work and family.

  As suggested by Quick, Henley, and Campbell Quick, there are three components using which one can manage oneself

---

[38] Thompson, C. A., & Prottas, D. J. (2006). Relationships among organizational family support, job autonomy, perceived control, and employee well-being. *Journal of Occupational Health Psychology, 10*(4), 100–118.

[39] Quick, J. D., Henley, A. B., & Quick, J. C. (2004). The Balancing Act—At work and at home. *Organizational Dynamics, 33*(4), 426–438.

well to maintain a balance in life in general: (a) managing work responsibilities, (b) managing family obligations, and (c) managing self-imposed expectations.[40] It is important that individuals have control over each one of these components, so that they can manage the pressure from all the domains more efficiently. People must realize that they play a vital role in defining the degree of work–life conflict they face.

WHAT CAN ORGANIZATIONS DO?

Maintaining employees' work–life balance is a shared responsibility of the employees and the employer. This is because a happy employee is an efficient and a more productive employee. On the contrary, if the employee is not happy, his/her productivity at work is jeopardized, which also affects the performance of the organization as a whole.

Kofodimos, in his work, indicated that the social character of individuals is responsible for the imbalances experienced.[41] Organizations should facilitate an environment and work culture that is conducive for work–life integration.

Work–Family Programs and Policies

Work–family programs and policies can play a major role in facilitating employees with better work–life balance. Thompson, Thomas, and Maier indicated in their study that the most responsive family programs and policies could be categorized into four categories: (a) dependent care, (b) parental leave programs, (c) spouse relocation and job locator, and (d) alternative work schedules.[42] Alternative work schedules would include facilities like flexible work hours, part time jobs, work from home option, etc. Generally, most of the policies and programs have been designed for employees who

---

[40] Ibid.
[41] Kofodimos, R. J. (1993). *Balancing act: How managers can integrate successful careers and fulfilling personal lives*. San Francisco, California: Jossey-Bass.
[42] Thompson, C. A., Thomas, C. C., & Maier, M. (1992). Work–family conflict: Reassessing corporate policies and initiatives In U. Sekaran & F. Leong (eds), *Women power: Managing in the times of demographic turbulence*. Newbury Park, CA: SAGE.

have kids and heavy family responsibilities. However, single couples or employees without families also derive benefits from this. It has been noted that mere access to the family policies and programs is of no good. It is important that the employees are satisfied with them. Rosin and Korabik pointed out that satisfaction with the family-friendly environment at work leads to a reduction in work–life conflict, and not just the mere existence of it. The family policies and programs should be in alignment with the individuals' needs.[43] Otherwise, organizational family-friendly policies and programs remain underutilized, and do not achieve their objectives. It is important for the firms to assess what their employees really need, and frame their policies accordingly.

Thompson identified four barriers in successful implementation of family-friendly organizational initiatives: (a) masculinization of work and feminization of the family, (b) lack of consensus and national leadership on work–family issues, (c) lack of support from supervisors and managers, (d) lack of evaluation of work–life programs initiated by the organization.

Corporate organizations need to work on the aforementioned barriers to make their policies more effective for improving their employees' work–life balance. Another study by Thompson and Beauvais exhibited that only 55 percent of the chief financial officers were of the opinion that employee's work–life balance was very important, and 39 percent of them said it was somewhat important. The study also showed that the most offered benefit to the employees was flexible work hours, with 45 percent of the organizations offering it. This was followed by part time work, job sharing facility, and telecommuting options, with 40 percent, 27 percent, and 17 percent of the organizations on the study offering these respectively.[44]

---

[43] Rosin, H. M., & Korabik, K. (2001). Do family friendly policies fulfill their promises? An investigation of their impact on work–family conflict and personal outcomes. In D. L. Nelson & R. J. Burke (eds), *Gender, work, stress, and health* (pp. 211–226). Washington DC: American Psychological Press.

[44] Thompson, C. A., Beauvais, L. L., & Lyness, K. S. (1999). When work–family benefits are not enough: The influence of work family culture on benefit utilization, organizational attachment, and work–family conflict. *Journal of Vocational Behavior, 54,* 392–415.

Managers in an organization play the key role in facilitating the family-related policies and programs to the employees. If the managers have a resistance to the family-friendly policies and programs, which they feel are detrimental to the firms overall productivity, they will make the policies inefficient. Thus, it is important that the managers are pro their organization's culture.

Often, organizational culture and its family-friendly policies clash. This can be explained with the fact that employees who use less family-friendly programs and policies have higher prospects for promotion, increment in salary, and incentives.[45]

Thompson, Thomas, and Maier proposed three dimensions of work–family culture: (a) managerial support for work–family balance, (b) fewer negative career consequences to be associated with the utilization of the work–family policies, (c) few organizational time demands that could interfere with family responsibilities. They also suggested from their study that employees utilized the family-friendly policies more, if the organizational culture was more supportive and vice versa. Organizations need to be supportive because such policies have a direct impact on the employees' commitment to the organization, increased retention rate, and less work–life conflict.[46]

Galinsky, in his study, found that individuals experienced less work–life conflicts when their managers and supervisors, and workplace culture were supportive. Also, it was noted that they had greater work satisfaction and were more willing to work harder, and help the organization attain higher productivity.[47] Kolb and Merrill-Sands identified a dual aim for organizations. *One was gender equity, and the other was increasing organizational effectiveness.*[48]

[45] Allen, T. D., & Russell, J. E. A. (1999). Parental leave of absence: Some not so family friendly implications. *Journal of Applied Social Psychology, 29*(1), 166–191.

[46] Thompson, C. A., Thomas, C. C., & Maier, M. (1992). Work–family conflict: Reassessing corporate policies and initiatives. In U. Sekaran & F. Leong (eds), *Woman power: Managing in times of demographic turbulence* (pp. 59–84). Newbury Park, CA: SAGE.

[47] Galinsky, E., Bond, J. T., & Friedman, D. E. (1996). The role of employers in addressing the needs of the employed parents. *Journal of Social Issues, 52*(3), 111, 136.

[48] Kolb, D. M., & Merril, Sands D. (1999). Waiting for outcomes: Anchoring a dual agenda for change of cultural assumptions. *Women in Management Review, 14*, 194–202.

Most of the organizations that have policies conducive to increase work–life balance are more successful in keeping their employees accommodated, rather than integrating their work and family lives. An organization should provide its female employees with unique policies that can allow them to perform efficiently at work, as well as fulfill their duties at the domestic end. Unique policies are required because women are exposed to situations that are more vulnerable to their careers. Pregnancy, recovery, taking care of children, etc., are inherently bestowed on women. Even though men try and replace these, there are some responsibilities that can only be fulfilled by women. Organizational environment that has no gender bias and designing policies keeping in mind the roles men and women play in a family, will establish work–life balance.

Rapoport said that work–life balance had been defined as keeping the work and the family domains away from each other. But he suggested that this definition should be changed and replaced with *work–personal life integration*.[49] Work–life balance meant adding equal importance to the work and personal life by looking at them in isolation. Organizations need to try and ensure that their policies and programs keep the employees satisfied with their work, as well as with their families.

An organization's work culture and ethics are the keys to provide a work–life balance to individuals. It is important to note that the employees do not have to make a choice between home and work; rather, they should be able to manage both. This is the only way an organization can have happy and efficient employees, who will, in return, increase the productivity of the organization, and, thus, profits.

Work–Life Balance in Military

Jacey Eckhart, a military columnist, (whose father was in air force, husband in navy, and son in army) shares her experience about work–life balance in the military.

---

[49] Rapoport, R., Bailyn, L., & Fletcher, J. K. (2002). *Beyond work, family, balance*. San Francisco: Jossey Bass.

Jacey Eckhart, the Director of Spouse and Family Programs at Military.com, in her article titled Balancing the military and family life says that there are two self-evident fundamental truths about work/life balance in the army (below is direct quote from Eckhart's article):
She says:

> In the military today, we seem to hold two truths to be self-evident about work–life balance.
>
> Truth #1 is that work–life balance does not and cannot exist in the military. I hear this belief from military spouses at live events all the time: *We are "all in." Army first. Family second. That's just the way it is.*
>
> Truth #2 is the belief that a little work–life balance is absolutely necessary. Spouses, I talk to, would love to have one of those husbands who carry exactly half of the child-care/house-care equation. We wouldn't mind bringing home an equal or better pay check. We could be in a family where everyone rides bikes, and eats lots of vegetables, and looks like Katherine Webb in an Alabama T-shirt, which would really be an interesting look for my husband.
>
> Personally, [quoting Eckhart] I don't think either of those "truths" work for military families. I don't think families can always come behind the military. We cannot live on scraps of time. I also don't think the military can suddenly start sending service members home from deployment, in order to attend every assembly at school. I don't think young Marines can confine their troubles to work hours. I don't think our ships or aircraft can just decide never to be broken.
>
> So, how is that supposed to ever come into balance? An interview with a business manager was conducted, who was formerly an army officer, about the steps to achieve work–life balance in the military.
>
> According to Eckhart the manager pointed out that the problem with work–life balance in the military is that we want to compare work and family. "There is no comparison," said the manager. "Family is first."
>
> Eckhart reports that the manager like he knew military members would throw back their heads and howl at that. As a West Point graduate and former artillery officer, the business manager was well versed in the fact that our military members do have to go when the military calls.
>
> That isn't the problem. Because balance is not achieved by parceling out time or divvying up hours. Instead, the business manager

emphasizes that balance is really about your attitude toward your family: *You have to want to go home reports Eckhart.*

Eckhart reports that the manager is right about that. Eckhart states that husband spends most of his time with the Navy. When he is home, his phone rings, and buzzes, and vibrates with Navy stuff. But we know he likes us best. We know he loves to be home. We know he works as fast as he can during work hours so that he can get back home.

Finally, Eckhart states that work–life balance in the military is not a scale where hours are weighed and measured. Work–life balance is not something that can be calibrated at a Jiffy Lube (car maintenance and service station). Instead, work–life balance comes from family first. We are the bedrock upon which our service member stands, constantly shifting to balance the demands of a military life. (This has been directly quoted from the document "Balancing the Military and Family Life" from the website www.military.com.)[50]

Another interesting case study of work/life conflict (below is direct quote from Dao's article) in army is of Specialist Alexis Hutchinson, a 21-year-old Army cook and single parent.

> Alexis was days from deploying to Afghanistan last fall when her mother refused to take care of her 10-month-old son for the duration of her one year. Army duty in a non-family station. Specialist Alexis's mother had a child of her own at home and attending to her sick sister, and working at the same time. Feeling overwhelmed, Alexis's mother took the boy back to Savanna, Georgia, where Alexis was posted, and begged her to find another person to care of her. That is when Alexis did what might seem natural to a parent, but to the Army it was a serious offense: she stayed home with her child and missed her flight to Afghanistan. She was arrested, and later charged with offenses that could have led to a court-martial and jail. On Thursday, Alexis received another-than-honorable discharge, ending an impasse between legal experts and army. Alexis's situation has started a spirited in military circles.... The legal wrangling over Alexis's case stirred much discussion on blogs, with

---

[50] Eckhart, Jacey. (2013). Balancing the military and family life. Retrieved from http://www.military.com/spouse/military-life/balancing-the-military-and-family-life.html (accessed on August 1, 2014).

sympathizers wondering why the Army would prosecute a parent struggling with child care problems, and critics questioning the soldier's motives. Alexis's mother has heard some of that criticism firsthand. Dao reports that Alexis's mother said "People have said to me, 'She signed this contract. She's supposed to go. That's her first priority,'" Mrs Hughes said, "My response is: 'I don't think so. This is her child. This is her family. This is her priority. The military is a job.'"

Several research studies have been conducted to understand the connection between work and no-work life among workers in the civilian sector. The research studies propound that balance between work and life effects constructs, such as psychological well-being, quality of sleep, substance use, and the quality of relationships with spouses and children. Therefore, it is important to consider the specific work–family circumstances of military families, and what strategies might best support their abilities to continue their military service without having to forgo family life, and reduce feelings of overload and strain, even when job demands are extreme.[51]

FAMILY WELFARE

By the Family, for the Family, of the Family

While the men are securing and protecting the borders for the nation, the wives and families are also tightly knit into the Military Systems. The Military System facilitates the integration of spouses and children of the men on active duty by programs such as AWWA.

AWWA or the Army Wives Welfare Association is an NGO run by volunteers who primarily comprise of families of the men on active duty. The primary purpose of such organizations is the development and well-being of war widows, army spouses along with children and differently abled.

(Box continued)

[51] Dao, James. (2010). Single Mother Is Spared Court-Martial. *The New York Times*. Retrieved from http://www.nytimes.com/2010/02/12/us/12awolmom.html?_r=0 (accessed on August 5, 2014).

(Box continued)

> The objectives of AWWA go a long way for the well-being of spouses and their children. Some of the objectives are as follows:
>
> a. Relief and rehabilitation of battle causalities and widows.
> b. Welfare of hospitalized patients.
> c. Provide support and vocational training to families.
> d. Run schools for differently abled children.
> e. Reward in the form of scholarship to meritorious and needy children.
>
> The guidelines are set by the President who is the wife of the Chief of Army Staff. Therefore the entire setup of AWWA is governed by the mantra "By the family, for the family, of the family."
>
> This work family integration is achieved through various programs such as:
>
> *Exceptional Family Member Program (EFMP)* is created to help the family members of the active duty personnel after the family get settled at a new place. This program constitutes essential articles, tools, resources educational trainings to the personnel family members.[52]
>
> *Military Spouses & Military Mom Speak* is an initiative by one of the major connection. Military Spouses make incredible sacrifices round the clock/365 days a year. Many have started pursuing their training for portable careers under MyCAA program too.[53]
>
> *The Career Resource Management Centre (CRMC)* provides guidance for searching employment opportunities for military

(Box continued)

---

[52] The Exceptional Family Member Program. http://www.militaryonesource.mil. [Online] http://www.militaryonesource.mil/efmp/overview?content_id=269174 (acessed on July 23, 2014).

[53] Military Connection—Military Spouses & Military Moms Speak. http://www.militaryconnection.com. [Online] http://www.militaryconnection.com/military-spouses-and-moms-speak.asp (acessed on July 23, 2014).

(Box continued)

> personnel family members and assist them in exploring career opportunities
>
> *Family Member Employment Assistance Program (FMEAP)* helps in achieving the personal and professional goals of the spouses by providing guidance counselling.[54]
>
> *Family Readiness Group (FRG)* is a sponsored organization of family members, soldier's volunteers and employees who are based in the camp. It provides support and assistance by creating a network for communication among the family members and those belonging to the unit. It is run by the FRG Leader who is generally a commander's spouse.[55]
>
> AWWA and many such programs are essential for ensuring the welfare and integration of families of the men on duty into the Army system. Such programs allow the army to serve the nation without any apprehensions about the well-being of their families.[56]
>
> Such programs were started by the soldier's families and spouses so that the spouses remain involved. These programs emphasise the importance of work–life integration and its deep impact on the success of our men on borders. Work, Life, and Integration are like the three legs of a tripod, which are together required to maintain a balance in life and without even one, this balance can be compromised.
>
> In the army, it is not about work–family balance, but it is more about work–family integration.

---

[54] Career Resource Management Center (CRMC). http://www.mccsmcrd.com. [Online] http://www.mccsmcrd.com/MarineAndFamilyPrograms/PersonalAndProfessionalDevelopment/CareerResourceManagementCenter/index.html (acessed on July 23, 2014).

[55] Family Readiness Group (FRG). http://www.fortbraggmwr.com. [Online] http://www.fortbraggmwr.com/acs/frg/ (acessed on July 23, 2014).

[56] About us. http://www.awwa.in. [Online] http://www.awwa.in/about-us.php (acessed on July 23, 2014).

8

ORGANIZATIONAL PRIDE AND UNITY

INTRODUCTION

Unity and pride are the two prime factors that keep the strongest and the most powerful military organizations in the world integrated and motivated. Unity is seen as a desirable feature of a group that leads to an increase in performance and productivity of the members of the military.[1]

Any organization flourishes because of its employees, who are dedicated to the cause of the organization. Any employee takes pride in both the job and the organization. A sense of belonging to the organization instills a positive working environment in the organization. In turn, it is also the responsibility of the organization to involve the employees in their goals and development projects. An organization would be more popular and would be taken pride in by the employees, if it is involved in some public centric activities, rather than the sole aim of making money. Any aware employee would find pride in working for an organization

---

[1] Oliver, L. W., Harman, J., Hoover, E., Hayes, S. M., & Pandhi, N. A. (1999). A quantitative integration of the military cohesion literature. *Military Psychology, 11(1)*, 57–83.

that belongs to the public, and, hence, is open to the general people, rather than the upper class. These employees would always remain with their parent organization than migrating to some other place for profits. The organizations understand the value of such loyal and hardworking people, and, hence, often carve out policies that are beneficial to both the employees and the company; since the employees, also, have to juggle between their family and social lives, and a disturbance in the family life would mean deterioration in the work quality of the employee at the organization.

Unity among the team members leads to unit cohesion in the military. It can be defined as "a dynamic process that is reflected in the tendency for a group to stick together and remain united in the pursuit of its goals and objectives."[2] It has also been well defined by Thomas Britt as "the holding together members of an organization to accomplish an objective by their commitment to each other, their organization, and organization's mission."[3]

A study conducted by Oliver, Harman, Hoover, Hayes, and Pandhi indicated that cohesion in the military is positively related to retention, well-being, and readiness. The study also indicated a negative relationship between cohesion and indiscipline. Oliver showed a positive relation between cohesion and combat effectiveness.[4] Based on various empirical studies, it has been concluded that high level of unit cohesion also has a positive relationship with *resilience* for combat-related stress. It has also been analyzed that unit cohesion *creates trust among the members* of a unit, and, also, its seniors/supervisors.[5] This has been referred to as horizontal versus vertical bonding. Trust has been regarded by many military leaders as the factor that *prevents emotional breakdown* in crucial times, and leads to *effective decision making and problem solving* in high threat situations.

[2] Carron, A. V., & Brawley, L. R. (2000). Cohesion: Conceptual and measurement issues. *Small Group Research, 31*(1), 71–88.

[3] Britt, Thomas (2005). Military life: The psychology of serving in peace and combat. *Military Performance, Praeger Security International, 1.*

[4] Oliver, M. (1990). *The politics of disablement.* Basingstoke: Macmillan.

[5] Griffith, J., & Vaitkus, M. (1999). Relating cohesion to stress, strain, disintegration, and performance: An organizing framework. *Military Psychology, 11*(1), 27–55.

Unity in the military keeps the unit *members motivated* and *perform efficiently under stress*. Unity in the military units prevents the soldiers from feeling alone. The onus of building and maintaining team cohesion lies on the leader and supervisors of the team. This is a way through which the military can mobilize capabilities of the individuals to achieve a common goal.[6]

Another important characteristic of unit/group/team cohesion is the *provision of support* to its members in case of difficult conditions. Another important characteristic of group cohesion in the military is resistance to external pressure, and being able to retain one's individuality through tough times.[7]

THEMES ON COHESION IN MILITARY

- **Task and Social Cohesion**
  Social cohesion in the military indicates cohesion based on interpersonal connections, and not on the basis of what they are doing. Task cohesion is a result of a single commitment to accomplish a particular task, which requires co-operation from the entire group.[8]
- **Primary Group: Locus of Unit Cohesion**
  Primary groups are groups that are small in size and interact face-to-face with each other. These groups prove to be the most powerful in group cohesion.[9] These groups have feelings of sympathy and mutual identification for their members. This group is complemented by a secondary group. In secondary groups, members are more widespread and have less face-to-face interaction.

---

[6] Winslow, Donna (1999). Rites of passage and group bonding in the Canadian airborne. *Armed Forces and Society, 25*(3), 429–457.

[7] Gross, N., & Martin, W. E. (1952). On group cohesiveness. *American Journal of Sociology, 52*(6), 546–554.

[8] MacCoun, Robert J., Kier, E., & Belkin, A. (2006). Does social cohesion determine motivation in combat?, *Armed Forces and Society, 32*(4), 646–654.

[9] Cooley, Charles Horton (1909). *Social organization: A study of the larger mind.* New York: C. Scribner's Sons.

- **The Four Components of Cohesion**
  The four components of cohesion are peer, leaders, organizational, and institutional bonding. This is also known as the *standard model for the US military* by various researchers.[10]

CREATING A SENSE OF MISSION

There is a strong sense of mission deployed in the members of the military in any country. The members of the military are in the military because they want to be a part of it. They are a part of the organization because their personal goals align with the goals and missions of the organization. Members of the military are aware of the struggles they would encounter throughout their work life, but are still willing to dedicate their lives to it.

The aforementioned sense of mission is found missing in the corporate organizations. Organizational mission and goals are not clearly communicated to the employees. A research, taken up by Campbell and Yeung, indicated that if an organization defined and communicated its mission well to the members of the organization, it can be managed better.

Like Peter Drucker said, "Business purpose and business mission are so rarely given adequate thought, is perhaps the most important cause of business frustration and failure." Organizational mission has not received an important place in the field of management. Many management thinkers talk of the organizational mission as an abstruse component of the organizations, some talk of it as a commercial discourse, and some also associate it with organizations' strength and identity. There are two ways in which the mission of an organization can be categorized into two approaches. One is mission in terms of a business strategy, and the second is, mission in terms of philosophy and ethics.[11] The first approach of mission, in terms of a business strategy, looks at mission as a strategic tool, which

---

[10] Kirke, C. (June 2010). *Military cohesion, culture and social psychology. Defense & Security Analysis, 26*(2), 143–159.

[11] Campbell, A., & Yeung, S. (1991). Creating a sense of mission. *Long Range Planning, 24*(4), 10–20.

will describe the organization's business rationale, target marker, and intellectual discipline. The second approach of organizational mission depicts the philosophy on which the organization is based, and the ethics it follows. The philosophy and the ethics approach of mission considers mission as a tool that performs the function of bringing collective unity in the organization. This brings strong norms and values that influence the way in which the members of the society behave.

Many organizations communicate the business strategy mission approach to their members, and not the second approach. However, without knowing and understanding the philosophy and the ethics of the organization, its employees cannot follow the business approach in accordance with the organization's expectations. This has been explained by Ted Levitt, a marketing professor at Harvard, in his article titled "Marketing Myopia." In this article, he indicated that many companies define their missions wrongly.[12] He gave an example of how oil producing companies depicted their missions. They depicted themselves as the oil manufacturing companies, and not as energy producing. He suggested that companies need to spend more time and thought in defining their missions.

Campbell and Yeung developed a model of mission for organizations on the basis of four prime elements (Figure 8.1). These four elements are *purpose, strategy, behavior standards, and values.*[13]

The *purpose* element of mission answers the questions: what does the company do and how did it come into existence? The purpose of a firm has been explained in the form of three categories. In the first category, the company claims that it is in existence for the benefit of a single group, for example, its shareholders. In this case, the purpose of the organization is to generate profits and wealth for its shareholders. However, this approach is single-minded and does not sound beneficial in a holistic sense. The second category of organizations is the one, which has the purpose of satisfying the needs of all its stakeholders. This would indicate working for the betterment of

---

[12] Theodore, Levitt (July 2004). Marketing myopia (HBR Classic), *Harvard Business Review*, Prod. No: R0407L-PDF-ENG, 138–149.

[13] Campbell, A., & Yeung, S. (1991). Creating a sense of mission. *Long Range Planning*, 24(4), 10–20.

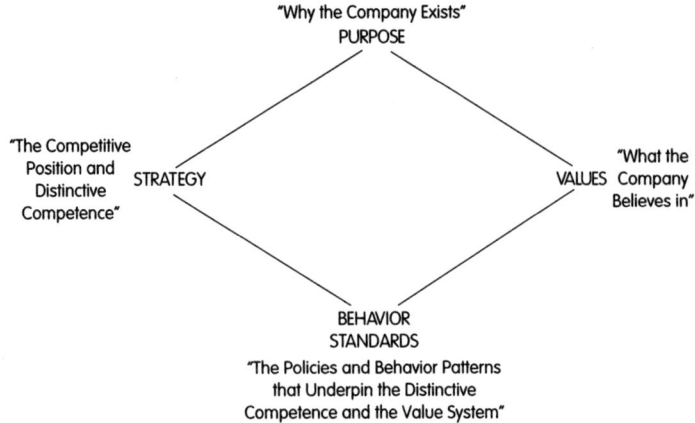

Figure 8.1 The Ashridge's Model Mission for Organizations

Source: Campbell, Andrew, & Yeung, Sally (1991). Creating a sense of mission, *Long Range Planning, 24*(4), 10–20.

its customers, employees, shareholders, etc. In the third category of companies, managers are not happy by just meeting the needs of the stakeholders. They want to be involved in a mission that has a greater a mission; a mission in which all the stakeholders come together and contribute to the society as a whole. There is pride associated with the kind of work they are doing. Here, everyone is treated equally, and all the stakeholders feel that they are doing a meaningful work.

The second element, *strategy,* is associated with the competing strategies that organizations need to follow to achieve success. Strategy of an organization is the commercial logic. So, if the purpose of an organization is to provide the best products then its strategy should define the principles on the basis of which the purpose will be attained. It will also lay its future plans and the unique strategy (product differentiation or cost advantage) it will follow to compete with the other firms.

The third element, *behavior standards,* are guidelines set to help people decide what they must do on a daily basis. Behavior standards act as a support for the purpose and the strategy an organization must adhere to. These behavior standards become identification for the employees, and mission, and values for the firm.

The fourth element, *values,* indicate the beliefs and the moral principles that shape an organization's cultural structure. The values and beliefs add meaning to the behavior standards. Values of an organization are often left unexplained. An organization should try to explain the perception of the philosophy and the ethics followed it. The aforementioned four elements together make a strong mission of an organization. Mission of an organization must be explained to the members of the organization. This develops an emotional bonding and commitment in the employees toward the organization. Behavior standards of an organization depict its mission to the members. Often, a company's mission aligns with the employees' values, which brings in them a sense of responsibility for their own actions. *People look for meaning and opportunity to transcend the ordinary. Values give meaning to the work employees do and make it fulfilling, as it has a greater purpose.*[14]

ORGANIZATIONAL CAPABILITIES

Organizational capabilities are the factors that make organizations more admirable and desirable. These abilities are collective skills, which are brought together by means of communication, training, etc. Organizational capabilities is about getting together all the available resources to accomplish the organizational goal and mission.

- **Employing the Right People**
  How does one define the right people? Placing the most intelligent notch of people in a team may not give the desired results. Different domains require different and specialized skill sets. Selecting the most capable and suitable candidate for a job is the right way to build a strong team. Employees feel united and carry pride in their organization when they know that the people employed are cautiously selected on the basis of their abilities, and that they all are a part of the organization, helping it in achieving its common goal.

[14] Ibid., 10–20.

Unity can be achieved when the team members have respect, empathy, and are supportive toward each other. When suitable candidates are appointed for a task, all the members know that they all are at par, and are, thus, equal. This feeling of equality even eliminates all the indifferences and incompetencies in the members, and develops the feeling of unity. Empathic, social, and supportive behaviors can be inculcated in the members of an organization by way of behavior standards.

- **Choosing the Right Talent**
  Choosing the people with the right talent can be explained as appointing competent people, who possess skills as per the business requirements. It is equally important that the appointed employees should be retained. Employees can be retained if leaders provide those who are contributing more with more of what they want.

- **Ensuring That Employees and Customers Have Positive Experience**
  It is important that the employees and customers of an organization have repeated positive experiences with the organization. As per an exercise undertaken by Harvard, organizations with 80–90 percent of the employees having a consensus about the organizations' goal performed better than the ones that had employees with a consensus of 50–60 percent.[15] It is important for the firms to know that their employees understand and are working toward their goals. To ensure that the customers have a positive experience, organizations should take regular feedbacks from them, and the relevant changes should be communicated and made to the employees.

- **Unity in Diversity**
  Today, all the companies have a diverse background of employees. People from different walks of life and different cultures are now dealing and working together in organizations. The work environment should be such that people from

[15] Ulrich, D., & Smallwood, N. (June 2004). *Capitalizing on capabilities*. Harvard: Harvard Business School Publishing Corporation.

different backgrounds can easily come together and unite into one team.

- **Focus on Common Goals and Likeness**
  One way of bringing unity among workers is by focusing on the common goals and likeness of the people. Unity comes when people find common interests. Common interests act as a connecting point for people, making the work environment more cordial and cohesive. Emphasizing on common goals and interests does not mean that individuals would not display their uniqueness. Rather, unity is about using the uniqueness of each individual in favor of the team and all the members in the organization. If one individual is better than the other in something then his ability can be used to help those who lack the same ability.

  Unity can be induced in an organization when employees accept that no one is perfect, and that everyone has their own weaknesses and strengths. Members should respect and appreciate others' strengths and help them in their weak times.

- **Make Employees Feel That They Are a Part of the Organization**
  People cannot perform to the best of their abilities if they think that they are not a part of the organization. Organizations must ensure that employees feel important, and their contribution to the firm should always be acknowledged. Employees can be treated as a part of the organization by making them actively participate in different meetings, seminars, etc. Also, everyone in the organization should be communicated about the recent developments in the firm. Feedbacks and suggestions should be taken from the employees, and the employees should be appreciated as well.

- **Showing Appreciation**
  Appreciation is a gesture that motivates people to perform better each time. It is a form of remuneration in kind. An employee, for making an achievement, should be appreciated in front of everyone. This motivates the employee to perform better, and it will inspire others to make more efforts. Generally, organizations make appreciations in the form of

financial remuneration. However, verbal and written appreciations prove to be stronger motivational tools and forms of appreciation. Mere financial appreciation (remuneration) lacks the ability to create sense of community, loyalty, and trust. It lacks a personal feel, and creates a competitive environment on the basis of the increase in wage among employees and companies. This reduces the retention rate in an organization, affecting the overall productivity and efficiency of the firm.

- **Integration**
One of the important determinants of the military's victory is integration among the members of the unit. Integrity holds all the officers together and makes them work as an efficient team, leaving behind their internal disputes. Integration is uniting the members of the unit at a professional level to achieve a common goal.

    Similarly, in case of a corporate organization, every employee, in some way, contributes to the organization he/she is working for. In a way, all of them are interlinked and connected to the common goal of the organization. Integrity is a factor that binds all the employees together. Workplace integrity is about creating and maintaining a healthy and respectable work environment. Such a workplace nurtures development of high professional standards, and demonstrates the values and ethics of the organization.

- **Credibility**
Organizations must build a sense of credibility in the employees by communicating with them on a regular basis about the new developments. This responsibility lies in the hands of the manager. The manager must ensure that he/she coordinates well with the available resources and employees to make the organizational goals clear. Credibility comes by converting communication into action. It is imperative that firms follow whatever they communicate.

    There can be no better organization than the military when we talk about an example of credibility. Military follows everything that makes the base of an organization. It follows everything it preaches. It teaches by way of being an example itself. The military first accomplishes its goals, and then motivates and directs others to achieve their personal and professional goals.

- **Fairness**
  Organizational fairness means treating all the employees equally. Employees should be given fair and equal amount of compensation financially, and in terms of benefit programs. Major decisions, such as promotion and salary increment, should be made without any bias, and purely on the basis of performance and merit. Everyone should be given an equal opportunity to perform well and for recognition. Fair and just environment in an organization makes the workplace healthy and competitive.

- **Foster Continuous Learning**
  Training your employees to be good performers needs a work environment, which promotes continuous learning. Continuous learning encourages the employees to learn more and perform better. It also urges them to take responsibility of their professional development and growth.

- **Shared Responsibility**
  Organizations must cultivate the sense of shared responsibility among the employees. Shared responsibility implies that every member of the organization has a responsibility to maintain the ethical and professional standards in the workplace. Shared responsibility connects the supervisors and the subordinates. This is because, in shared responsibility, a supervisor is equally responsible and accountable for his/her subordinates' performance and behavior. This builds integrity among the members of the organization, leading them to support and cooperate with each other.

- **Opportunities on the Basis of Merit**
  The world has been plagued with social unrest because of various types of discrimination. It is certain that such discrimination would only create unrest among the members. Most of the companies, today, have understood this, and have implemented policies that prevent biases or discrimination on the basis of gender, caste, religion, etc. However, problems

may still exist if a supervisor or anyone has any bias toward anyone in the organization. Such issues should be dealt with utmost care and sensitivity to make the workplace desirable for people coming from anywhere.

Employees should be given equal opportunities on the basis of their performance and merit, irrespective of their caste, gender, etc. This ensures that the right talent is employed in the organization, assuring its increasing productivity and growth. Leaders play an important role in uniting the members by acting as role models.

- **Controlling with Penalties**

    Penalties/punishment is a controlling measure that encourages the members of the military to abide by its values and principles. In the military, the supervisor/commander has the duty of assessing the situation in case of a conflict. He has to undertake sessions with various parties to the incident, such as the accused, witnesses, and victims, and, accordingly, give punishments. The commander has the power to punish the miscreants with suspension, salary cut, physical exertion, etc. The commander has a lot of power, but it is important that he uses it in the wisest way possible. His action will leave a print on others' minds, and can also prevent them from repeating mistakes made by the guilty.

    In the corporate world, punishments cannot be given in the aforementioned military style. However, similar impact can be made on the members in corporate organizations. This can be done by clearly communicating the organizational values and behavioral standards to the employees. They should be informed and warned about the repercussions of their unacceptable acts as per the organizational norms.

    Employees should know that their acts can lead them to pay penalties in different forms: salaries can be deducted, employment can be terminated, etc. It is important that these practices are practiced strictly and fairly. All the members should be treated with the same penalties. This will develop trust in the employees for the organization they are working for, and will motivate them to work harder and prove themselves.

- **Men and Women: At Par**
  The socio-economic scenario has changed significantly over the past few decades. Women are now treated equally in every sector and domain. They have proved themselves time and again. However, it has been observed that not many women occupy the top positions in firms. There are many reasons for this. First, could be that, the peak period of a woman's career clashes with her optimal reproduction period. Most of the times, women are compelled to leave work and give priority to domestic affairs.

  Women in the military were not given an entry to the combat team a couple of years ago. Now, new appointment systems have been adopted to ensure that men and women are chosen on the basis of their ability and merit, even in the combat teams. In case of the navy, it was indicated that more or equal number of women improved the condition of the men in the navy. When less or no women travelled with the navy officers, they were often found drunk and in a miserable state. However, as more women entered the field, men had to maintain themselves and a certain amount of decorum too. This prevented them from going overboard in a drunken state, and also from other bad habits.[16] Also, lack of women in the field led to problems like physical assault, misbehavior with women, etc. This clearly indicated that more women in the navy helped the men in maintaining the right state of mind and the environment as a whole.

- **Understand Employees' Space**
  Understanding the space of employees can increase their willingness to work and perform better. Understanding the private space of the employees is to understand that people have a life of their own, outside their offices and professional lives. Organizations should ensure that work does not interfere with the private space of their employees. Intrusion of work into the private space of individuals leads to a conflict between the two most important domains of an individual's life, work

[16] Capt. Abrashoff, D. M. (May 2002). *It's your ship—Management techniques from the best damn ship in the navy*. New York, USA: Business Plus, Hechette Book Group.

and home. Studies have indicated that these conflicts affect the well-being of the individual, affecting his performance levels at work and at the home front.

LINKING ORGANIZATIONAL PRIDE WITH ORGANIZATIONAL SUCCESS

Pride is an important factor contributing to the success of an organization, as well as the productivity of its employees. Pride is the prime factor that makes the members of the military strive to make an entry in the organization and survive through the training period. Pride in the members of the organization makes them perform better, increases their retention, and keeps them motivated. Pride in the military is associated with the mission and the values of the organization. People join the military with the knowledge that life is not going to be simple. Officers have to deal with rigorous training, staying away from their families, witnessing traumatizing incidents in battlefields, etc., but they carry on with their duties and responsibilities.

Similarly, organizations need to build pride amongst their members. Pride can be built in the employees by communicating to them the company's mission, and their importance and contributions to the firm. Pride is a stronger emotion than job satisfaction. A recent research indicated that employees do not have a positive attitude toward their employers. The study indicated that 59 percent of the employees were not sure of where their organizations were heading toward. Forty-two percent of the employees felt that their employers shared poor communication strings with them. Only 48 percent of the employees surveyed said something that would encourage their friends to join their employers. About 37 percent of the employees said that they would leave the organization if they were not proud of working for it.[17]

A study indicated that the pride in being employed by an organization can affect employees' recommendation to others to work with the organization, and recommendation to use its products and

---

[17] Anonymous. (September 2004). Pride before profit: A review of the factors affecting employee pride and engagement. *A CHA Report*. Retrieved from http://zookri.com/Portals/6/reports/cha-report04-pride.pdf

services. It also influences the extra effort that employees put in, and the activeness with which one looks for a job elsewhere.[18] All of these behaviors have a substantial impact on the firm's performance.

PRIDE-ACHIEVEMENT RELATIONSHIP

Harter defines pride as, "an emotional response to an evaluation of one's competence."[19] Pride is no more a positive emotion. It is an emotion that reflects the accomplishments of an individual. Therefore, to bring pride in the employees, it is important that they are regularly updated with their accomplishments and must be rewarded for the same. Organizations must communicate and inflict through their actions and gestures to the employees, the importance of their contribution to the organization.

An acknowledgement of achievement acts as a motivating tool for the employees, which makes them realize their self-worth. Therefore, to build pride, you should make the employees feel proud of themselves.

BUILDING PRIDE

There are various factors that can build pride in the employees about the employer. Employers must make sure that they treat all the employees properly. This indicates an organization's culture and values. An organization should also be known for its contribution to the society and the community. It should be able to make difference to people's lives. An organization's customer care also adds to its reputation. If the customers are happy, they will look up to the organization with high regards. It is important that organizations that make good products are also innovative. Organizations that take

---

[18] Interbrand. (2007). Who cares whether they care? A study of the effect of pride in organizations. Retrieved from http://webcache.googleusercontent.com/search?q=cache:WMK352bIt1wJ:www.docstoc.com/docs/32468319/Pride-Before-Profit+&cd=1&hl=en&ct=clnk&gl=in (accessed on August 6, 2014).

[19] Hatter, S. (1985). Competence as a dimension of self-evaluation: Towards a comprehensive model of self-worth. In R.L. Leahy (ed.), *The development of the self* (pp. 55–121). New York: Academic Press.

initiatives in saving and protecting the world environment gain more respect from the employees and people in general. Organizations need to take initiatives and strive to obtain better results for themselves, and society in general. Organizations get what they give.

Factors that Influence the Feeling of Pride for an Organization

As per a study by InterBrand, the following are the factors with the highest correlation with boosting pride for an organization in the order of ranking.[20]

- Organization's products and services are known to be the best.
- Organization has high standards of work.
- Organization is well known for addressing its customers.
- Organization treats its employees with respect.
- Organization has inspiring and supportive managers.
- Organization gives people the opportunity and responsibility to make a difference.
- Organization has a positive impact on people's lives.
- Organization pays fairly well compared to its competitors.
- Organization is positively represented in the media.
- Organization has high ethical standards.
- Organization has been consistent in producing good financial results.
- Organization contributes to the community welfare.
- Organization has celebrated heritage

Organizations should make efforts to work on all the aforementioned factors. Building pride in the employees will be a holistic process. Pride is certainly an important factor that can reduce employee turnover and increase employee productivity. Proud employees also

---

[20] Anonymous. (2007). Who cares whether they care?. InterBrand as cited in Donna Dickson (2008) "Fostering employee engagement: A Critical competency for hospitality industry managers." Retrieved from http://scholarworks.rit.edu/cgi/viewcontent.cgi?article=1685&context=other) (accessed on August 5, 2014).

act as ambassadors for the company they work for. It not only has a positive effect on the employees, but also carries its positivity in the form of increased shareholder value, more satisfied customers, and increased profitability. An organization's job is not just creating good products and services, but it also appoints, retains, and encourages the best minds to perform the best work possible. To make this possible, it is important that organizations have positive, empowering, and supportive cultures. Employees are a firm's greatest assets.

BUILDING ORGANIZATIONAL PRIDE THROUGH CORPORATE CULTURE

Corporate culture is that component of organizational behavior that can synergize unity and pride in the members of an organization. Corporate culture is basically a set of guidelines adopted by an organization, describing its values and beliefs. As per Ouchi, corporate culture is a tool that brings about harmonious professional relationship among the co-workers.[21] The values and beliefs of the organization should be enforced and communicated to others by way of policies and informal norms. When the co-workers are guided by the culture of an organization, it could bring enthusiasm to the work force, thus bringing greater commitment to work, increased productivity, and increased profits.[22] Corporate culture also brings self-identity to the individuals. The culture followed in an organization becomes a roadmap for its members to follow. This then becomes the moral code of the organization.[23] Corporate culture brings unity and pride among the members, as it motivates them to do things together, for each other, and for the organization as a team.

[21] Ouchi, W. G. (1980). Markets bureaucrats and clan. *Administrative Science Quarterly*, 25(2), 129–141.
[22] Martin, J., & Frost, P. (1996). The organizational cultural war games: A struggle for intellectual dominance. In S. Clegg, C. Hardy, & W. R. Nord (eds), *Handbook of organization studies* (pp. 599–621). London and Thousand Oaks: SAGE.
[23] Fineman, S. (1999). Emotion and organizing. In S. Clegg & C. Hardy (eds), *Studying organizations: Theory and method*. London: SAGE.

All Work and No Play, Makes Jack a Dull Boy

The aforementioned statement holds true when the work culture is such that employees get no time to have fun. Fun does not only mean in terms of provision of recreation facilities to the employees, but it also means bringing a fun element in work. Making work a fun thing to do totally depends on the culture of the organization. There are organizations who believe that no fun should be allowed while working.[24] The question is, if employees are not enjoying whatever they are doing, for how long will they want to do it, and how efficiently will they do it. Lack of fun and enjoyment factor make work monotonous and mundane. It reduces the effort that employees put in to improve their performance, and, hence, can reduce productivity. An organization does well if its employees are happy with the organization they work for, and with the work they do.

Army Organization and Unity

Being in army requires a person to be always in unity with his team. "Cohesion is a bond of relationships and motivational factors that make a team want to stay and work together."[25]

Based on their own experiences, the company commanders (working under various armies) in the Company Command forum have been talking about the best, most cohesive teams they have served in, as well as about how they foster cohesion in their units.[26] These stories have a lot to learn from about the various aspects of an organization. A few of these experiences from these commanders have been quoted directly.

---

[24] Abrashoff, D. M. (2002). *It's your ship: Management techniques from the best damn ship in the Navy*. New York, USA: Warner Books.

[25] U. S. Army (2012). Army doctrine reference publication 6–22: Army leadership. United States Government.

[26] Anonymous. (April 2013). Building a cohesive team. Company command. Retrieved on July 28, 2013 from http://webcache.googleusercontent.com/search?q=cache:k686laznR_sJ:cc.army.mil/pubs/armymagazine/docs/2013/CC_ARMY_(Apr2013)_Cohesive_Team.pdf+&cd=1&hl=en&ct=clnk&gl=in

- **Ari Martyn**

    Three things stand out about the cohesive unit I joined fresh out of Ranger School:
    I believed that my unit had higher standards than other units. That is to say, it appeared to me that the unit held itself to a standard not found in sister units. I found out later that the commander had a role in getting this started, but eventually it just became a self-fulfilling prophecy. The leadership knew their Soldiers well—not in an intrusive way, nor in an inappropriately friendly way, but in a professional yet personal way. I was impressed with how much the leadership knew about every single Soldier, his habits, his background, embarrassing stories, etc. Also, there were a lot of nicknames, but they all were positive or funny, not demeaning.

    There was a strong, robust welcome program. For example, I was picked up at the airport in Italy by the commander and his wife on the weekend. The senior PL [platoon leader] was waiting at the company for me, where he took me to the hotel room they had booked for me, and then took me for an orientation tour of the post and the city. That night, all the officers had dinner together. The following Monday, one of the PLs took me over to in-processing and then to draw my equipment. And this same model was copied at lower levels, for example, for an arriving squad leader, and even, to an extent, for the privates. Their welcome ritual was not hazing; rather, it was expressed as "Welcome to the team. We have tough standards, but we're going to set you up for success because we are all going to war together on the same team."

- **Scott Safer**

    I am on a Security Force Assistance Team deploying in 30 days to Afghanistan. Team cohesion is paramount to our 13-men organization. We sat down with the officers this past week and discussed this topic. One of my lieutenants wrote down his thoughts:

    "In my experience, shared hardship is the most effective way to obtain team cohesion. Every individual has his "wall" he needs to overcome, a moment, when the route looks too horrid to continue. It is when a person reaches this point that his true character is revealed. The same goes for a team. If the members of a team reach this point and discover unification, they will overcome the obstacle and conquer many challenges that may come their way. If the team crumbles, it will reveal a weakness that needs to be addressed. By enduring this common struggle, teams share experiences that cannot be compared to other relationships. The team members begin to know each other's weaknesses and strengths in a variety of situations, learning how to utilize these attributes for the next task. Once a group conquers a shared hardship through teamwork, its members develop mutual trust and confidence in one another. A leader who conducts rigorous training, provides team-building exercises, and accomplishes challenging missions will develop a cohesive team. During these events, however, the leader needs to instill a positive attitude and share the burden of the task. If subordinates witness their leaders not undergoing the same hardships, or are not given a reasonable purpose for the task, such events may cause an adverse reaction within the team."

- **Andre Fields**

    The most cohesive team I have ever been on was in the National Guard. I knew I was on a cohesive team when Soldiers in the unit did *everything* together. We were like an extremely close family! Even though the unit was deactivated six years ago, we still get together as a team at least once a year. It all started with the commander. He fostered respect and loyalty within the team, and we all knew he had our backs no matter what, with no hesitation. Once we saw this, we extended that respect and loyalty back to the commander, and it permeated throughout the team. This is what I have fostered in my own company as a commander. It has taken the unit from being in shambles to a very cohesive unit—a family.

The Darker Side of Cohesion

Cohesion: Most of us presume it to be a positive commitment to the mission and to all soldiers in the team. But imagine a tightly bonded group whose values and norms run counter to the army's. Is that possible?

Some of the commanders shared their views on the darker side of cohesion, which are:

- **Pete Exline**

    "While deployed, I used a unit survey to help identify one platoon that seemed cohesive on the outside, but actually had bitter trust issues. My first sergeant brought in a new platoon sergeant—a rather quiet and soft-spoken NCO who, along with the right mix of strong E-6s, did a great job of restoring trust and cohesion to the platoon."

- **Sam Linn**

  "Extremely high cohesion can be great for morale and may make a yearlong deployment more palatable at the tactical level, but it can encourage group thinking and stifle diversity of ideas. In my experience, there is often an 'out' group that is not being heard, and can make the group difficult to work with for outside agencies and adjacent units."

- **Josh Christy**

  "If cohesion comes at the expense of original thinking, creates turf wars and is destructive to relationships then it is not the cohesion we want. Cohesion is not a goal in and of itself; it must serve the mission and foster trust for everyone."

- **Pete Kilner**

  A team is too cohesive if its Soldiers prioritize their loyalty to each other above their loyalty to Army values. Such a team risks covering up unethical behavior and dealing with it solely "in house." Leaders must ensure that cohesive teams are as loyal to our professional values as they are to each other.

  Unity reinforces commitment—to each other, to the unit, and to the mission. And this commitment to the people, the team, and to the cause is what brings out the best in a person. If a person feels that he has many hands to support him when in need of help, he becomes dedicated to the cause and the organization. Army has been the best possible example of an organization, which has developed this unity in its people. A soldier is the basic unit of an army, and both army and the soldier thrive because they are proud of each other.

(Note: These have been directly quoted from the document Building a Cohesive Team. Company Command from the website cc.army.mil)[27]

[27] Anonymous. (April 2013). Building a cohesive team. Company command. Retrieved on July 28, 2013 from http://cc.army.mil/pubs/armymagazine/docs/2013/CC_ARMY_(Apr2013)_Cohesive_Team.pdf

> **PRIDE AND TRADITION: LEARNING FROM THE PARA**
>
> The Parachute Regiment in Indian Army was formed on March 1, 1945 including four Battalions and four independent companies. On April 15, 1945, the Parachute Regiment was re-raised by absorbing the following three erstwhile British era Parachute Battalions. 1st Battalion the Punjab Regiment, 2nd Battalion, the Maratha Light Infantry, 3rd Battalion, the Kumaon Regiment. PARA and PARA (SF) battalions in Parachute Regiment in the Indian Army consist of exclusive volunteers. In India all the Special Forces personnel are chosen among the Paras and they are commonly known as "Commandos." The main motto of the parachute regiment is "Shatrujeet" that is the "Conqueror."[28]
>
> The main aim of having a Parachute regiment battalion is for quick deployment of the soldiers behind enemy lines to attack the enemy from behind and destroy their first line of defence. In order to accomplish the operational tasks, the manpower selected in the regiment should be young, physically fit, mentally stout, innovative, and should be highly motivated.[29]
>
> The goal of the para units is to function as an elite battle unit of the infantry. Paratroopers specialize in secret activities geared to disrupt enemy operations. Furthermore, paratroopers are trained to carry out operations in vital areas in enemy territory with an element of surprise. Para Commandos are supposed to achieve the following goals:
>
> - To create commando superiority in a battle area
> - To interrupt the operations of the enemy stealthy
>
> *(Box continued)*

[28] Editorial Team. Parachute regiment, paratroopers of Indian Army. http://www.ssbcrack.com. [Online] http://www.ssbcrack.com/2013/12/paratroopers-of-indian-army.html (acessed on July 23, 2014).

[29] The parachute regiment. http://www.indianparachuteregiment.kar.nic.in. [Online] http://www.indianparachuteregiment.kar.nic.in/welfare.htm (acessed on July 23, 2014).

(Box continued)

- To disrupt the communication channels of the enemies
- To weaken the enemy major areas and points[30]

The three parachute commandos (battalion-size units) perform Special Forces duties. Airborne, Air Assault or Parachute troops are usually held centralized. The mounts, in all cases, are provided by the Indian Air Force. All the parachute regiment battalions classified as Special Forces, are considered to be reaction troops. However, the use of parachute regiment forces is dependent on the air transport fleet.

The selection process for the parachute regiment entails that all officers apply either as "volunteers" fresh from recruitment or transfer into para-units from regular army units. Parachute battalions maintain a probationary period of ninety days for the applicants.

Each of the applicants is assigned a probationary officer. The role of the probationary officer is commendable as he is the one who turns volunteers into commandos. Under the mentorship of the probationary officer, these volunteers undergo a six months training in the Special Forces Regiment and three months training for a normal parachute regiment. The training includes mental, physical and psychological tests by their assigned probationary officer. Team work and camaraderie are critical for success in clearing the probation. This probation period assists the officer in charge to understand the volunteer's objectives and goals and map them to those of the army. The applicants who successfully finish their training program can become Parachute Commandos within the Army. The selected applicants are welcomed and a "Maroon beret" is presented to mark them as paratroopers.

The aforementioned details indicate how probation period can be best utilized to recruit the best. Further, how army traditions are used to gauge person–organization fit.

[30] Para commandos. http://www.bharat-rakshak.com. [Online] http://www.bharat-rakshak.com/LAND-FORCES/ParaCommandos.html (acessed on July 23, 2014).

ABOUT THE AUTHOR

Dr Dheeraj Sharma is a faculty in Indian Institute of Management, Ahmedabad, India. He earned his doctoral degree with major in marketing, and double minor in psychology and quantitative analysis from Louisiana Tech University, USA. He has taught and presented research at numerous education institutions in North America, Europe, and Asia.

Dr Sharma has been the recipient of many academic and professional awards. He is an American Marketing Association Doctoral Consortium Fellow and National Conference of Sales Management Doctoral Fellow. He was nominated for Clifford J. Robson Excellence in Teaching Award, and also for Erica and Arnold Rogers Excellence in Research Award for year 2009–2010. He was an associate editor of *Journal of Marketing Channels,* and also an editor of the Academy of Marketing Science proceedings. He is an active member of Academy of Marketing Science, Society of Marketing Advances, American Marketing Association, and World Marketing Congress. As a member of the aforementioned associations, he has served as a session chair, discussant, track chair, and executive member of organization committee.

Dr Sharma has over fifty publications in leading international journals, encyclopedias, books, and conference proceedings.

Dr Sharma has been involved in consulting projects and executive training with several multinational corporations, such as ICI Paints,

ABOUT THE AUTHOR

Duncan Holdings LLC, Globe Rangers LLC to name a few. He served on board of India Canada Cultural and Heritage Association, Canada. He continues to be a special invitee to Army Management Studies Board of Indian Army.